Benthic Foraminiferal Ecology

Patrícia Pinheiro Beck Eichler
Christofer Paul Barker

# Benthic Foraminiferal Ecology

## Indicators of Environmental Impacts

Springer

Patrícia Pinheiro Beck Eichler
Federal University of Rio Grande do Norte
Natal, Rio Grande do Norte, Brazil

Christofer Paul Barker
Ocean-land Interaction
EcoLogicProject
Boulder Creek, CA, USA

ISBN 978-3-030-61462-1 ISBN 978-3-030-61463-8 (eBook)
https://doi.org/10.1007/978-3-030-61463-8



This Springer imprint is published by the registered company Springer Nature Switzerland AG
The registered company address is: Gewerbestrasse 11, 6330 Cham, Switzerland

# Preface

Imagine the joy of a child innocently playing in the mud. Go ahead, seriously, please take a moment, perhaps close your eyes, breathe easy, and imagine… MUD! Wonderful mud! It is one thing that connects humanity at its greatest innocence. Every child everywhere loves mud! In this moment of our imagining, she is surrounded by an inner and outer world of fantasy, wonder, and beauty. Like a dream, this world is as real to this bright and curious child in that moment, as this world is to you and I in this very moment.

Time passes and now this little girl is all grown up. Among her life's many great accomplishments, she has become a mother, a mentor, an internationally recognized research scholar, and professor of Biological Oceanographic and Paleoclimatology.

I am honored and proud to introduce my best friend and greatest inspiration, Dr. Patricia P.B. Eichler. She, like her own mother before her, is teaching her children and students the wonders of what is really happening in that mud.

So, what is happening down in the marine sediment, and why is it so important? Single-celled microorganisms called Foraminifera live there! Benthic Foraminifera specifically, or "Forams" as they are affectionately referred to. That is what is happening. Forams are so incredibly fascinating and so important that they have become her hobby, career, and life's passion because their many differing species provide insights into our past, present, and future with empirically accurate data.

Not just any data, but climate science data that shows earths preserved geologically impacted change patterns preserved in their shells occurring from millions of years to this moment. Data that reaches over time to help us predict the future! A future where logic and reason shows us that pollution and extinction are not parallel lines, and that eventually these lines will meet; and that timeline appears to be in rapid approach.

Here we hope to impart the wisdom that Foraminiferal ecological data related to abiotic patterns gives us, with their use as stable isotopes for micropaleontology and paleoclimate studies. As cofounders of EcoLogicProject.com our endeavor is to provide accurate data for calibration of the present and past and share these predictable indicators for ocean warming and acidification, as well as sea level rise and loss of biodiversity. We demonstrate how climate change may affect future changes, which include sea level rise in a warming and acidifying ocean, generating cyclones for example, and including environmental, social, and economic loss. Among the many things we propose, we also wish to make sound recommendations what we hope will "dissuade our collective persistent march toward mass extinction."

Rio Grande do Norte, Brazil — Patrícia Pinheiro Beck Eichler
Boulder Creek, CA, USA — Christofer Paul Barker

# Acknowledgments

The authors are grateful to the Marine Geology and Geophysical and Environmental Monitoring Laboratories (Laboratório de Geologia e Geofísica Marinha e Monitoramento Ambiental Coastal/GGEMMA) and to the Graduate Program in Geology and Geodynamics (PPGG) of the Federal University of Rio Grande do Norte (UFRN).

We very much appreciate the ongoing support of Higher Education Personnel Research Coordination (Coordenadoria de Aperfeiçoamento a Pesquisa de Pessoal de Nível Superior/CAPES) and their many funded projects, starting with Dr. Eichler's master fellowship in 1994, and continuing to the present day. Through the Ocean Sciences Project (Edital Ciências do Mar II N°43/2013, Project N°23038.004320/2014-11), Postdoc Fellowships at UFRN, Brazil, was possible from 2011 to 2016. We also acknowledge the Special Visiting Professor (Professor Visitante Especial) Fellowship (PVE 151/2012 and AUX PE 242/2013) in the Science without Frontiers Program of CAPES, including a laboratory engagement at the University of California Museum of Paleontology in Berkeley and a speaking engagement at the Geology Sciences of Stanford University in Palo Alto California (USA) in 2015. We also thank PROBRAL 337/10 (CAPES/DAAD) for postdoc grants (N° 8116-12-1) at Kiel University in Germany in 2012. This research would not have been possible without the abovementioned funding of CAPES (Ciências do Mar II N°23038.004320/2014-11) and the postdoctoral fellowship grants (N°88887.305531/2018-00, N°88881.188496/2018-01, N°9999.000098/2017-05), at Moss Landing Marine Laboratories at San Jose State University (MLML/SJSU), at the Ocean Sciences Department of the University of California at Santa Cruz (UCSC), and at the International Ocean Discovery Program (IODP) through Texas A&M University (USA) in 2018.

We want to thank the Technical Support to Strengthen National Palaeontology (Apoio Técnico para Fortalecimento da Paleontologia Nacional, Ministério da Ciência e Tecnologia MCTI/National Research Council CNPq N° 23/2011, N° 552976/2011-3) where the "biodiversity of sensible ecosystems in the coastal area of reefs" in northeastern Brazil was studied and verified the possibility of recommending a biodiversity conservation area with practical applicability of science.

The career path success of our chief scientist and lead writer, Professor Eichler, would never have been possible without the incentive and emotional support of Dr. Helenice Vital. She is truly a "vital" person for the marine geology and geophysics of the northeast region of Brazil. Her diligence and accurate research improve and validate the lives of all around her. This includes her students, technical supporters, and many other collaborators, who depend directly upon her for their respective research. She is one of the most competent researchers and scientists of our time and deserves to be honored with humble gratitude and deep appreciation. With her manifold abilities to deal with multidisciplinary research, and to respond accurately to the most adverse situations, she has driven herself and her team to success in many different career paths. These include mentoring geologists, geophysics, geographers, physical, chemical, biological and geological oceanographers, biologists, marine biologists, divers, technicians, engineers, and mathematicians. All of whom have successfully accomplished their own research, studies, and endeavors. She unites people by sharing resources and workload in balanced terms, at the same time gently educating so we all have felt both adequate and a sense of belonging.

We would like to give special acknowledgment to Dr. Moab Praxedes Gomes, also a wonderful mentor, having been influenced by his mentor Dr. Vital, with the understanding that dedication and leadership are values and qualities benefiting both team members and team leaders. Dr. Gomes describes the ocean bottom features with accuracy, curiosity, methodology, and enlightenment that have led us to discover the wonderful world of paleochannels and submerged rivers in the sedimentary and coral reefs of the Rio Grande do Norte state.

Part of the field and laboratory work illustrated in this book from estuaries, bays, and mangroves in the coastal zones of the South, Southwest, and Northeast Brazil was financially supported by the São Paulo State Research Support Foundation (Fundação de Amparo à Pesquisa do Estado de São Paulo, Brazil/FAPESP Procs. 96/4191-8; 98/05409-2; 99/10678-5), the Oceanographic Institute of São Paulo University (IOUSP) in Brazil from 1994 to 2004, in the micropaleontology laboratory of Dr. Beatriz Beck Eichler, and the National Research Council (CNPq). Besides field and laboratory work, we are also grateful for the postdoc fellowship from the Delaware Geological Survey (DGS) and from the College of Marine Studies (CMS) of the University of Delaware from 2004 to 2009, which provided the understanding of marine stable isotopes and hypoxia and acidification in coastal environments specially enclosed Bays.

Thank you to João Pildervasser of Life Sciences of Springer Nature. His well-mannered approach, knowledgeable research specific to "Foraminifera," and understanding of scientific papers brought Dr. Eichler this invitation to share this work—a wonderful subject that has been the focus of her career and research, as well as that of her own mother before her.

As an international research biologist, oceanographer, and lifelong dedicated ecologist, Patricia is deeply concerned about the collective human impact effects to our beautiful planet. In her words: "I am deeply grateful to Christofer Paul Barker and the opportunity to co-promote the EcoLogicProject ideal". Our shared vision is to illuminate the message of eco-stewardship and preservation through environmental sustainability. Like myself, Christofer has dedicated his life to this endeavor. His love for the environment, and all life, has brought him to discover the scientific "y" side of his observations and correlations, including research on biodiversity and sustainability to preserve and conserve nature now and for future generations. For these and many reasons, I have invited Chris to coauthor this book, as we further our collaboration in bringing the message that a better world will exist when people with similar environmental ideals and interests unite for "the preservation of the environment and its biodiversity in its prime stage." Chris says; "he could not have said it any better describing Dr. Eichler, or in reciprocation of appreciation, for her meteoric knowledge and understanding, kindness, patience, wisdom and tireless dedication."

"My reason to exist," says Patricia, "are my daughters Maria Gabriela and Anna Julia, which my life is dedicated to. They show me that patient guidance and love results in a formula for success. I hope to share this message so that others also prioritize responsibility for sustaining and preserving a world that will support their lives, and their children's lives. For that, above all, I am thankful for the teachings and example that I receive from my own parents every day. Bea a loving naturalist and Otto a brilliant engineer. Both have dedicated their lives to improve my life and the lives of all around us. With the example of love and care, they have taught me to look upward and work for a better future, to live in the present, and learn from the past."

Lastly and perhaps, most importantly we both wish to thank the reader for doing the work and taking the steps to preserve, conserve, and connect with nature. It is that connection that roots us firmly in the tree of life, as part of, not apart from, Mother Nature and her infinite bounty that supports and sustains all life.

EcoLogicProject.com

Patrícia Pinheiro Beck Eichler Christofer Paul Barker

# Introduction

We will show you how Benthic Foraminifera are the most powerful climate and ecology tool that scientists can use as environmental indicators. They are the "First Responder Indicator Species" and are readily found in marine sediments worldwide showing us geomorphology which identify potential problems.

Research and academia often operate under the constant pressure of ongoing high-dollar perpetual funding. Other added bonuses that Forams provide are that they are an accessible, inexpensive, easy-to-handle proxy of gargantuan statistical significance.

These "tiny giants of the great seas" earn that title in humble yet esteemed scholarly research circles. The microscopic shells of these creatures are made of calcium carbonate ($CaCO_3$) and the marine sediment layers provide us with information of what has been happening with the carbon record over time. Each widely distributed specific Foraminiferal indicator species differentiates particular habitats (and geohabitats) from deep oceans to brackish lagoons, estuaries, and even freshwater. Fossils from these tiny Foraminifera yield clues about past climates.

As "the first responders" these plentiful unicellular microorganisms react rapidly to the slightest environmental changes. Their shells carry a record of these alterations, showing the absolute environmental condition of habitats and ecosystems that the creature lives in at the moment of its sampling. Because their shells also fossilize, they show us an accurate historical record pattern into paleo ecosystem conditions. This makes it possible to accurately identify climate and condition patterns from millions of years past up to our own unfolding recent environmental patterns.

To quantitatively measure environmental impacts, foraminiferal communities' dynamics are used simply because their rapid response to change is so graphic that it is easy to read. Abundances of key foraminiferal species generate comparable index values temporally or spatially. Besides the value of being distributed globally on land and in virtually all marine sediments, foraminifera provide yet another benefit as a model of "calcifying symbiosis" for testing stressors that threaten the ecological integrity of coral reef ecosystems. Data on benthic foraminiferal absolute and relative abundances or indices from marine sediment are applied worldwide to quantify changes in benthic ecosystems under natural and anthropogenic influences.

Benthic foraminiferal communities' dynamics as environmental impact indicators effectively identify and classify patterns, as well as differentiating natural processes, or those caused by human interference. Their distribution pattern applicability, in different kinds of environments on coasts all over the world, from pristine undisturbed natural environments, to heavily polluted and/or contaminated effluents, helps researchers identify problems and provide recommendations for mitigation and remediation solutions.

A considerable part of the beauty, accuracy, and simplicity of foraminiferal research is that although we need academic scholars and their interpretations, a great deal of this research simply requires minimal technology applied by technicians with modest training. Moreover, there is even ample near-future opportunity to implement artificial intelligence should academia feel the need for more expensive and even faster techniques.

# Contents

# About the Authors

**Patrícia Pinheiro Beck Eichler** Our Chief Scientist, Academic Advisor, and Cofounder Professor Eichler is a multi-linguist and mentors undergraduate, master's, and PhD students at private and public universities and schools internationally. She gives students a positive and welcoming experience to help them thrive in their science training. This includes the opportunity to boost their careers as coauthors within her extensive publications including journals, articles, press releases, etc.

Professor Eichler is among the world's foremost scientists in the biological oceanographic environment, with specialized experience in coastal processes, communities dynamics, sediment transport, and estuarine dynamics.

**Christofer Paul Barker** Cofounder and Operations Coordinator, Christofer is passionate about EcoLogicProject's mentorship and consultancy ability to accurately forecast climate change effects and provide disaster mitigation and remediation solutions. His dedication to environmental sustainability and vision of bringing fact-based climate science forward have coalesced into collaboration with Dr. Patricia Eichler and her international oceanic climatological research. Land-ocean-atmosphere interactions and mechanisms research allows him to also contribute as editor in numerous science publications and provide scientific technical support for the public, students, and client's services and needs.

His consultancy services include eco-stewardship solutions for viable sustainable nontoxic alternatives, leading to The EcoLogicProject Partnership.

# Chapter 1
# Playing in the Mud: Benthic Foraminiferal Communities—Dynamic Environmental Impact Indicators

**Abstract** We show the applicability of the use of foraminifera communities dynamics correlated with abiotic parameters to provide data in environmental quality and coastal management plans. In this sense, we pinpoint sites most subjected to pollution and contamination and we identify differentiate effluents even when salinity and temperature values form gradient. Foraminiferal assemblages show increase in abundance of tolerant species to specific abiotic parameters. Density and diversity negatively correlated with heavy metal and PAHs. Increase in percentages of deformed or abnormal organisms and reduced size of the organism when they are exposed to heavy pollution mainly due to Hg, PAHs, and PCBs. It was observed that tolerant opportunistic species (*Ammonia tepida* and *Buliminella elegantissima*) benefit directly from organic contamination increasing their relative abundance. The other type of benefit may be indirect, as has occurred for *Elphidium* spp., which greatly increases its dominance due to the absence of other species that have had their shells dissolved first, resulting primarily from the impact of the oil spills. On the sedimentary and coral reef areas, *Amphistegina gibbosa* tests were not found at sites where reefs are walked upon. The presence of hard substrates and coarse sediment, rather than water depth, seems to be controlling the distribution of the tests. The results to date reaffirm the use of foraminiferal species in studies of oil spills, contamination by sewage, and industrial pollution even where the dissolution of carbonate is active due to low pH values. We then conclude that pollution or contamination of an environment overlaps the "natural" zonation and gradient of environmental factors, therefore limiting the establishment of species that are strategists and not opportunists.

**Keywords** Pollution · Contamination · Foraminifera · Thecamoebian · Opportunists · Strategists · Abiotic environments · Interface sediment-water



P. P. Beck Eichler, C. P. Barker, *Benthic Foraminiferal Ecology*,
https://doi.org/10.1007/978-3-030-61463-8_1

## Introduction

Our goal is to provide readable reliable accurate information, for anyone to understand, about environmental factors that influence patterns of benthic foraminiferal distribution in transitional and marine ecosystems. In addition, we intend to identify patterns or potential problems now or that may arise in the near future, and to make sound recommendations toward sustainable 'eco-stewardship' for people to act upon, adapt, and make necessary mitigation and remediation changes.

Benthic foraminifera are environmental bio-indicators, especially in polluted and contaminated environments where their sensitivity to stressors are expressed by their assemblages, to alterations of the communities' dynamics. Their distribution patterns in ecological bio indicator pathway studies show up as early as 1959, with Zalesny studying the Santa Monica Bay in California, USA. He stated that 'environmental factors such as currents activity, nutrients, salinity, characteristics of bottom sediments, and especially temperature should control the distribution patterns of the living benthic species in that bay. Since then several works have focused on the effects of various types and sources of pollution in different marginal habitats (Setty and Nigam 1982; Sharifi et al. 1991, Alve 1991, 1995). The number of these kind of studies has significantly increased according to an extensive review of research over the last decade about Foraminifera as bio-indicators of pollution (Suokhrie et al. 2017).

## *What "the Hell on Earth" Is Going On?*

The problem is human; more accurately, the attenuating western colonization lifestyle model, where greed trumps reason and accountability, and our 'anthropic' activity has caused negative impact effects worldwide, for the post-industrial age, and also for the millennia!

Industrialization contaminants and pollutants increase from domestic or industrial sources are discharged daily, into our air, land, coastal areas and oceans. This measurable increase has accelerated greatly since the early 1960s. Humans in mass are out of touch with 'living as caretakers gently integrated with nature'. Global pollution from industries and domestic contaminations negatively affect everything that sustains life—water, air and soil.

The planet earth, our mother-earth-nature, literally provides everything we need to thrive. The pursuit for "more stuff", power, and of course money, while polluting our life-source is illogical and unsustainable. The effects not only pose human health threats, but also initiate eventual mass extinction of many other species, including our own. Our eco-stewardship is to protect the source of life, not abuse it. Otherwise, we will leave our children and the following generations cesspools of toxic and acidic wastewater—or worse. At some point, somewhere, we all came from "indigenous people". Indigenous people have always sustainably protected natural resources, looking ahead for seven generations into perpetuity. The reason why we are all here now, it is because indigenous people have always done this.

Persistent pollutants like polycyclic aromatic hydrocarbons (PAHs), polychlorinated biphenyls (PCBs), pesticides, heavy metals and others have a particularly deleterious effect everywhere, including to aquatic life. Organic matter contamination and accumulation from domestic runoff is another problem. Large amounts of untreated sewage also flow directly into lakes, rivers, bays, estuaries, mangroves and oceans worldwide. Untreated sewage carries virus and bacteria, which transmit deadly diseases such as cholera, hepatitis and the new deadly coronaviruses. Unfortunately, the sewage effluents often will deposit closer to our houses, and yet we may only experience the damaging effects much later on. Large amounts of organic matter dumped in coastal areas also cause algae 'blooms' proliferating and releasing toxins that kill aquatic life. The environment suffers from oxygen depletion from fast algae blooms, and saline and fresh water mass density differences with high stratification can also enhance stagnation, preventing mixing of water of different densities, worsening effects. Pollutants and contaminants enter the food chain as well, and the effect on the "first responders and consumers" shows up reaching the highest trophic levels.

## *What Are We Going to Do?*

The solution, in part, is to identify natural from anthropogenic patterns, and reverse the effects of environmentally negative human activity. This sounds easy but it is not. This profoundly complex problem can, however, literally be solved with a very simple approach.

We need to use the conditioning and programming effects that education, media, and religion already provide, to use these media channels, to reconnect humanity with Mother Nature the source of our life, and with our eco-stewardship. It is that simple! From home to kindergarten, all the way through higher education and beyond, we can use curriculum, theocracy, and multimedia to influence everyone.

Through the cooperation of international private and intergovernmental multimedia and religious-leadership, we can influence each other to create a better narrative. One that provides solutions oriented messages, and to promote awareness, self response-ability, volunteerism, accountability, and actions that sustain, preserve and maintain our ecology.

In the meantime, at least some definitive answers and solutions for coastal and marine environments are found right now through researching simple microorganisms in mud. After all, they are the first consumers at the lowest level of the food chain, and the first to respond showing positive or negative effects in the ecological chain of "natural" or "disturbed" environments!!

## *How Are We Going to Do It?*

Good question… Well, we are doing our best to inspire everyone to do their best. This includes creating resources and services that optimize our EcoLogicProject message, encourage participation, and influence positive outcomes.

Foraminifera are inexpensive easy-to-handle environmental indicator solutions. They help us identify contamination and pollution sources. They help us rapidly summarize environmental characteristics on and offshore. They highlight environmental variations over short periods, with sensitive reactions to seasonal changes and 'anthropogenic' effects. In addition, another great advantage is their abundance and our ability to easily collect them.

Some Forams species respond favorably to pollution and become dominant in contaminated areas, while others react negatively, decreasing in size, or becoming absent. In polluted areas, benthic Forams usually decrease in diversity until only one single opportunistic species remains.

Globally foraminifera establish themselves in natural, polluted and contaminated conditions, giving us the 'clues to the scene of the crime', 'who done it', and 'how to solve the case'.

Brazil is a prime example for at-risk environmental 'crimes-against-nature' conditions with mounting crisis looming. Our studies here show areas experiencing hazardous pollution or contamination. Forams clearly show us these problems, and target the offending sources, which allows us to highlight solutions.

Here in this first chapter, our intensive case study focus has stretched over 2400 miles (3870 km) along the coast from Laguna in the south, north up past Rio de Janeiro and Bahia, ending at Natal. We are looking specifically in semi-enclosed lagoons, bays, estuarine channel systems, and coral reefs where Forams unravel the environmental issues these coastal areas have.

Below we see first responses that foraminiferal associations have to different ecosystems. We show problems and information about geomorphology and indicator species, and provide recommendations to dissuade the persistent "collective march" toward mass extinction in this our Anthropocene era within these wet systems.

## *Laguna, Santa Catarina: A Semi Enclosed Coastal Lagoon (with the World's Only Wild Dolphins That Cooperatively Help Humans Catch Fish!)*

Located 130 km south of Florianópolis (capital of Santa Catarina state) are three interconnected lagoons: Santo Antônio, Imaruí and Mirim, which are about 40 km long. We started with 25 biological and geological samples at the entrance of the lagoons near Laguna city, followed by sampling inside these lagoons, until reaching the entrance of D'Una River (Imbituba city) (Fig. 1.1).

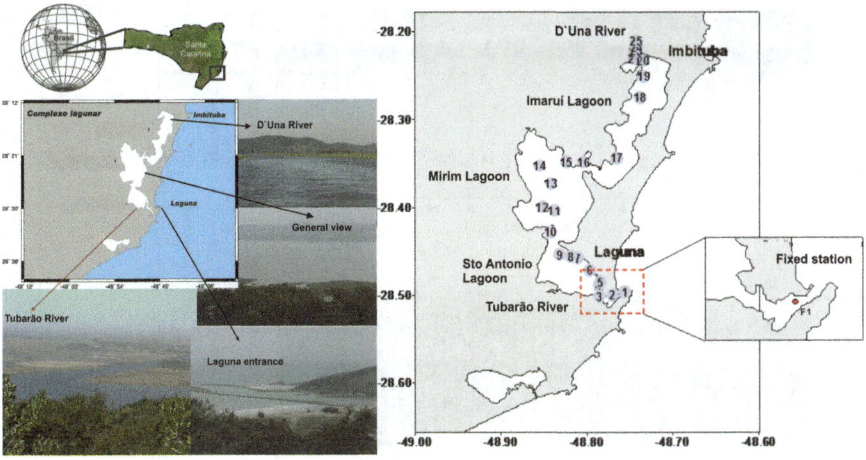


**Fig. 1.1** (**a**) Study area. (**b**) Marine sediment samples and measurements of hydrographic properties with samples collected in summer and winter and fixed samples (F1) in summer

This lagoonal system is characterized by elliptical cells connected to the adjacent ocean by a single narrow channel, which allows classifying it as suffocated. They are mesotidal lagoons with high-energy gravity waves. The channel acts as a dynamic filter, and consequently effects of tidal and current oscillation are attenuated inside the lagoon.

This type of lagoon has large fluvial discharge times, dominant wind effect, and intermittent variation of vertical stratification due to solar heating and freshwater discharge (Miranda et al. 2002). There are marsh grasses on the lagoons' margins, and this area is the southern limit of mangrove occurrence in Brazil. The risk potential in this area deals with the chemical industry to the north of the lagoon close to D'Una River, the use of illegal mesh in nets for fishing, and site constructions to the south close to Tubarão River. Close to urban areas of Imbituba and Laguna, there are untreated domestic sewage and contaminating aquaculture dumping practices in places where Bottlenose Dolphins (*Tursiops truncatus*) live and have an interspecies cooperative interaction with fishermen. However these amazing dolphins are slowly going extinct in part because of illegal mesh fishing net ('ghost nets') practices that are threatening their populations.

Our salinity and temperature study in these lagoons revealed colder temperatures in the bottom waters when values are compared to the surface waters, and salinity variation is intense in winter (Fig. 1.2a, b).

Descriptive univariate data show decrease of biodiversity revealed by the total number of benthic foraminiferal species toward inner parts of the lagoons. Moreover, in a clear type of zonation, their shells features are a response to the "natural" salinity and temperature gradient in the environment provided by the interface sediment-water where they live. Forams shells can be calcareous (made of $CaCO_3$) occurring in the entrance of the lagoon, and agglutinated (attaching sediment grains to its shell) occurring towards middle and inner part of the lagoons (Fig. 1.3a, b).

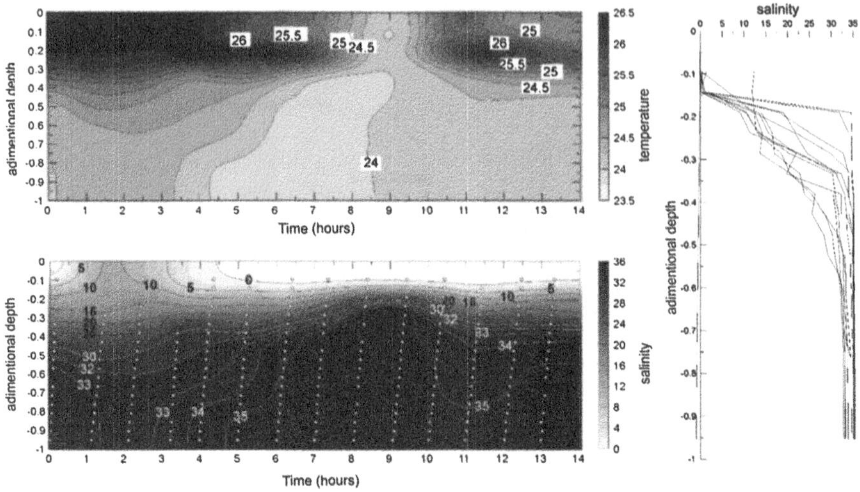


**Fig. 1.2** (**a**) Temperature and salinity in the fixed station. (**b**) Vertical salinity profile in the fixed station

**Fig. 1.3** Basic differentiation of shells ((**a**) Calcareous; (**b**) Agglutinated)

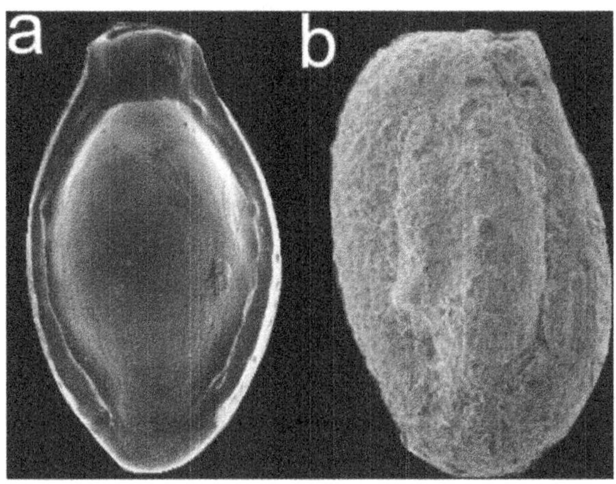


Besides descriptive univariate analysis in biological data to verify biodiversity, we also applied a multivariate approach in abiotic data using Principal Component Analysis (PCA) and a Multi-Dimensional Scaling (MDS) in the biological data. PCA was applied to the abiotic data matrix in winter and summer to explain the variability of lagoons' samples. PCA (Fig. 1.4a, b) shows that summer **(AV, CV, DV, BV)** and winter groups (**AI, BI, CI, DI**) are similar and respond to similar abiotic parameters (depth, $CaCO_3$, temperature, salinity, oxygen, surface pH, granules, sand, silt, clay, and organic matter).

MDS in matrix of summer biological data show formation of four groups (Fig. 1.5a), and winter biological data shows samples presenting two groups

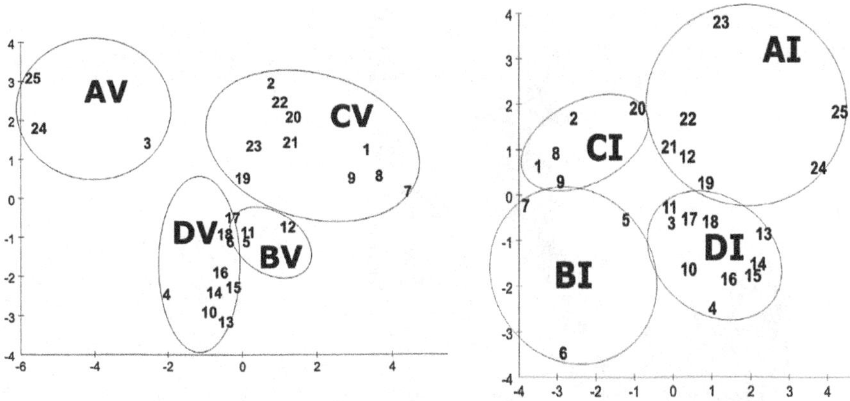


**Fig. 1.4** (**a**) PCA in the matrix of hydrographic and sedimentological data of summer (where AV: depth; BV: temperature, salinity, surface and bottom oxygen and superficial pH; CV: deep pH, granules and sand; DV: silt, clay, organic matter and $CaCO_3$). (**b**) PCA in the matrix of hydrographic and sedimentological data of winter (where AI: depth and $CaCO_3$; BI: temperature, salinity, oxygen, superficial, and bottom pH, and granules; IC: sand; DI: silt, clay, organic matter)

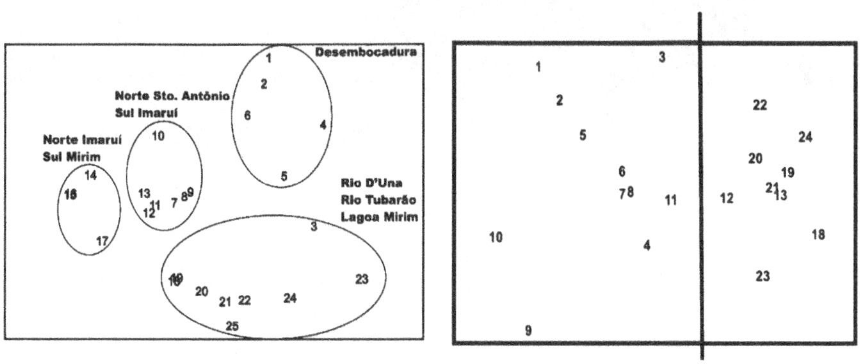


**Fig. 1.5** (**a**, **b**) MDS analysis in the biological data matrix in summer and winter

(Fig. 1.5b). This setting also show similar samples in relation to salinity and temperature zonation.

This salinity and temperature gradient is remarkable and foraminiferal communities derived from it can be observed in Fig. 1.6 where a zonation map with main dominant foraminiferal and thecamoebians (microorganisms indicative of higher freshwater input) species are distributed along a salinity/temperature gradient.

Foraminiferal associations present well-defined spatial distribution and zonation resulting from environmental conditions established due to seasonality and tidal influence of continental input and marine waters. Near the ocean, succession of marine influence and mixohaline environments present calcareous species (*Quinqueloculina* spp. 19–21; *Saccamina sphaera* 18; *Cassidulina subglobosa* 17; *Pseudononion atlanticum* 16; *Buliminella elegantissima* 15; *Bolivina striatula* 14;

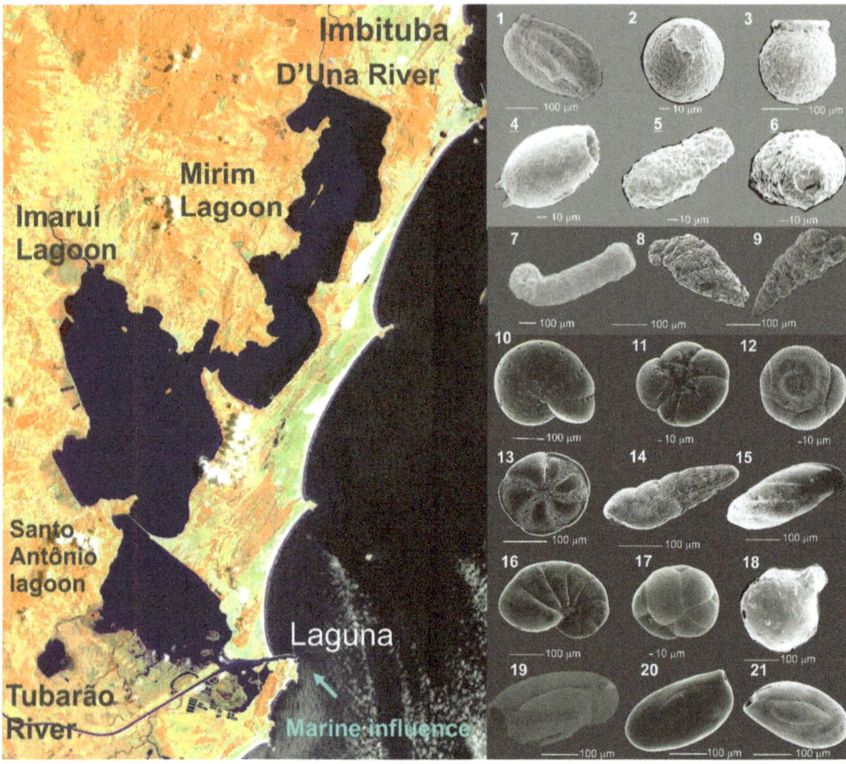


**Fig. 1.6** Zonation map with main dominant foraminiferal and thecamoebians species

*Buccella peruviana* 13; *Ammonia* spp. 12; 11; *Elphidium poeyanum* 10). Mixohaline environments with calcareous and agglutinated species (*Gaudryina exillis* 9; *Ammotium salsum* 8; *Ammobaculites exigus* 7) in the central part. In the northern part of the lagoons, where fresh water input is more intense, the agglutinated *Miliammina fusca* (1) and thecamoebians: *Pontigulasia compressa* (6), *Difflugia capreolata* (5), *Centropyxis aculeate* (4), and *Difflugia oblonga* (3, 2) are indicative of higher freshwater input.

This significant horizontal gradient observed along the lagoons is the result of balance between the periodic intrusions of saline waters into the system by tide and fresh water input from Tubarão and D'Una rivers. Those two rivers are important because of the urbanization of the cities of Tubarão, Laguna and Imbituba in the state of Santa Catarina, Brazil. Tubarão River banks have coal and thermal power plants, and D'Una River waters contain waste from rice monoculture entering the water flow daily. Tubarão River is located closer to the entrance and is more subject to the renewal of its waters by the ocean. The D'Una River is located very far from the ocean and its waters are not benefited with the renewal of ocean waters. Moreover, due to the higher depth of this river (5 and 7 m) in relation to the lagoon complex (less than 3.6 m), the circulation and oxygenation of the deeper waters is highly compromised.

This data indicates decrease in dissolved oxygen content in the water showing less mixing and therefore the sediment of the north region of the lagoon tends to be seriously impacted from human activity and resulting acidification.

Low oxygenation and salinities in the innermost parts of the Mirim lagoon, close to entrances of D'Una and Tubarão rivers are habitat for fauna, and patterns are related to the double influence of continental input and saline water intake, as indicated by salinity and distance from the ocean. This double influence has been described in many parts of the world and is the most prominent feature found in Brazilian coastal areas.

Forams population density patterns reflect environmental conditions and the groups representing north of Imaruí lagoon and south of Mirim lagoon have foraminifera and thecamoebians due to accumulation of silt, clay, and organic matter evidenced in the analyses of PCA. At the entrance, the samples most influenced by marine waters exhibit the highest diversities, while north of Santo Antônio Lagoon, and in southern part of Imaruí lagoon, agglutinated species dominate. Groups formed by samples collected in the D'Una and Tubarão rivers, and Mirim lagoon are similar with agglutinated *Miliammina fusca* and thecamoebians species, considered indicators of inefficient water renewal.

The low species establishment in inner sites of lagoons are related to contents of organic matter, silt, clay, and salinity, since most of these species do not tolerate low salinity levels. Closer to the ocean, species diversity and individuals' numbers increases, indicating that salinity controls species distributions. In winter, temperature decreases more than 10 °C in some samples, and it is responsible for decrease in number of foraminifera and thecamoebians, limiting distribution of these organisms. No evidence of eutrophication of the environment or even any type of degradation or other forms of pollution is observed. The tendency of accumulation of organic matter, silt and clay in the narrowing of lagoons does not yet present an alarming situation, but the little renewal capacity of waters is decisive in accumulation of pollutant or contaminant materials.

Pollution dumped in D'Una River tends to accumulate in inner parts and it is not flushed out of the system. Tubarão and D'Una rivers are subjected daily to deleterious pollutant products, and foraminiferal distribution shows possibility of serious problems in the near future, even though it is located closer to the ocean.

The faunal distribution pattern showed above is characteristic of the environment, showing early signs of human activities of pollution or contamination from coal plants, thermoelectric plants, and rice monoculture acidifying the sediment. Harmful effects to the environment from urban development are noticed in northern areas close to D'Una River, in central parts where renewal of waters is inefficient, and at Tubarão River entrance. Since the concentration of pollutants is high in this river, if no coastal management plans for sewage and industrial effluents and environmental monitoring are implemented in the near future, the environment will move towards degradation and collapse. This threat will surely affect the already threatened bottlenose dolphin who cooperatively assist fishermen to catch fish from shore at this specific site, a phenomenon only found in two places globally; here and a few kilometers to the south.

## *Bertioga Estuarine Channel Is a Tide-Dominated Estuarine Channel in Sao Paulo*

Moving northward, the Bertioga Estuarine Channel is located within the complex called Baixada Santista, on the south coast of São Paulo, and includes the municipalities of São Vicente, Santos, Cubatão, Guarujá, and Praia Grande (Fig. 1.7). It is divided into bodies of water, mountains, mangroves, sandbanks, Atlantic forest, beaches, rivers, estuarine and urbanized areas. Bertioga Channel, a secondary connection from the ocean to the Santos estuarine complex has two entrances. The northern entrance is located close to Bertioga city, and the southern entrance is close to Santos Channel, which flows into Santos Bay, a very polluted and contaminated site. Bertioga Channel is 25 km long, and among the rivers flowing into this channel are the Itapanhaú, Caiubura, Iriri, Tia Maria, Cabuçu and Trindade. Bertioga Channel is a low energy region and a mixing zone between marine and fresh water with relatively high temperatures allowing for the development of mangroves. Such environments are subject to variations in the abiotic parameters, where fluctuations of salinity and temperature are associated with tides.

Results from 88 samples collected from six distinct monitored stations monthly from April of 1999 to March of 2000 indicate differentiation of environments related to balance of fresh and marine water influence. These environments have conditions for the establishment of specific dominant species, from each different environment: brackish, mixohaline and euhaline. Figure 1.8 shows a differentiation map

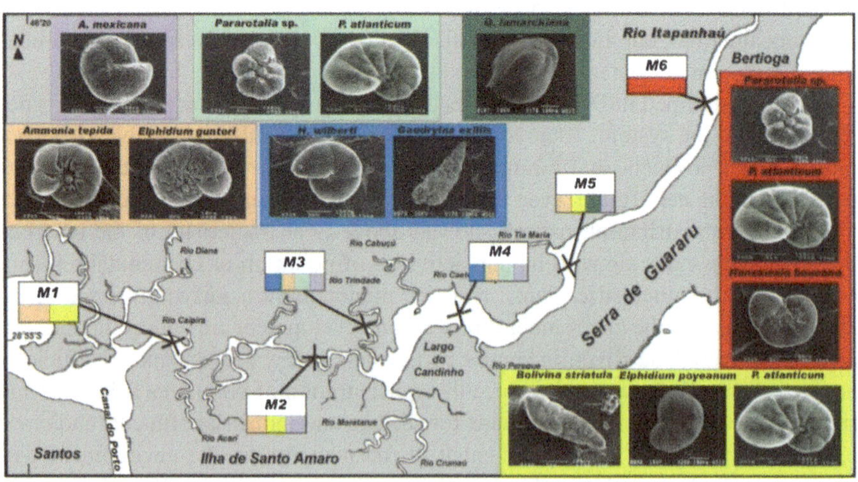


Fig. 1.8 Differentiation map with main foraminiferal species found in the area

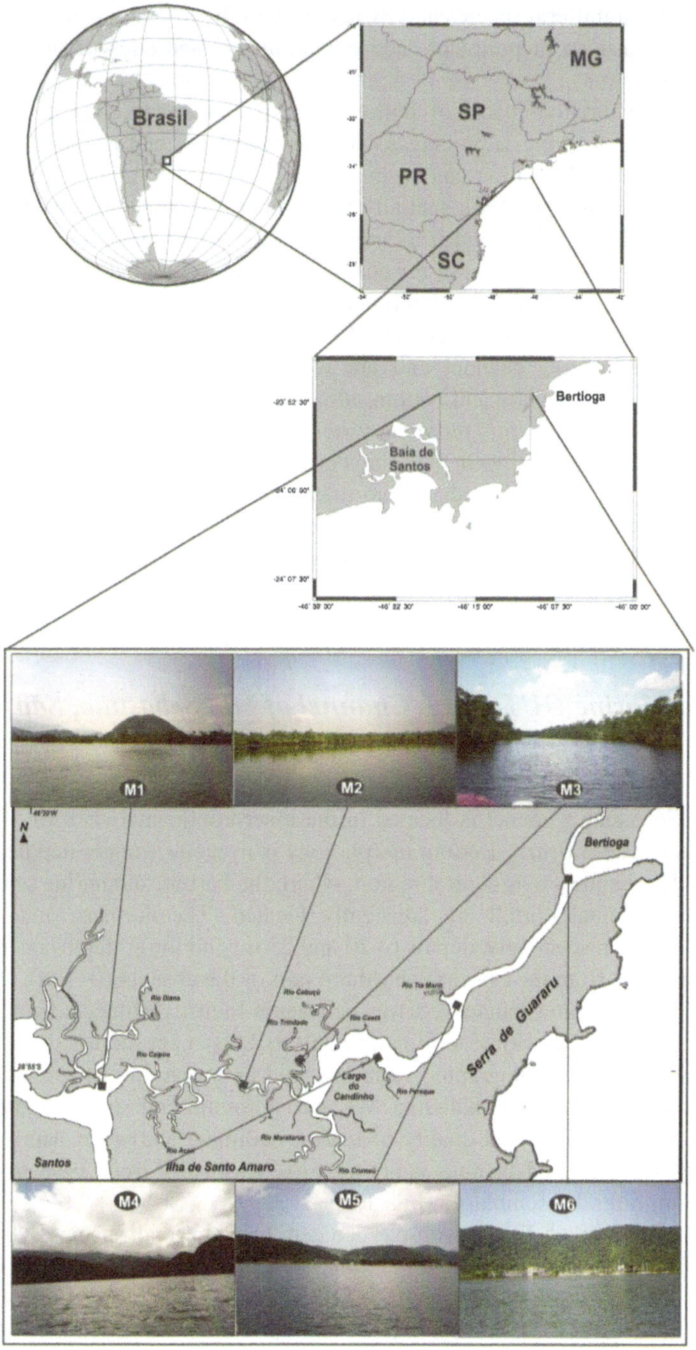


**Fig. 1.7** Tide-dominated estuarine Channel (Bertioga, SP) and its monitoring stations

with main foraminiferal species and its response to the hydrographic condition, its relation to transport of pollutants and contaminants, and sites prone to develop worse scenarios.

Monitoring direct impacts of pollutants and contamination on foraminiferal fauna are based on living assemblages where its abundance and species diversity are related to water circulation and sedimentation patterns. *Ammonia tepida* and *Elphidium* spp. were dominant spatially and temporally. Agglutinated species such as *Arenoparrella mexicana* and *Haplophragmoides wilberti* were found in brackish and mixohaline inner portions of the channel, mainly in Largo do Candinho and in the south (next to the Santos channel) and are linked to low values of diversity and evenness. *Pararotalia cananeiaensis* and *Pseudononion atlanticum* were found in both entrances of the channel and are indicative of higher currents velocity. Difference between surface and bottom salinity is related to increased in dominance of *Pararotalia* sp., indicating turbulent water dynamics.

Opportunistic species live in the south entrance close to impacted areas of Santos Channel, and marine species inhabits north entrance, with mixohaline species in the inner portions of the channel, which has predominance of muddy sediments with large amounts of organic carbon. The south of this channel is clearly vulnerable in its southern portion.

## *The SubMarine OUT FALL Channel of São Sebastião, São Paulo*

The São Sebastião Channel is located in the inner continental shelf of São Paulo State, southeastern Brazil. Bottom morphology is irregular, greater depths (>45 m) occur in the channels axis central region, where the harbor, submarine sewage and petroleum terminal outfall are adjacently located. The northern and southern entrances have, respectively, depths of 20 and 25 m, and lower depths occur on the continental side from the central to northern part of the channel (0–7 m).

The São Sebastião Channel receives petroleum industry effluents from TEBAR (Almirante Barroso Maritime terminal, PETROBRÁS). Considering that this channel is subjected to oceanographic variations and anthropic influences, monitoring the submarine outfall is a challenge. We used foraminiferal species indicators to determine vulnerability and quality of marine sediment and water using 81 sediment samples collected in nine sampling trips seasonally from 1998 to 2002 to study foraminiferal assemblages (Fig. 1.9).

Large numbers of living well-oxygenated foraminiferal species, young and adults of *Quinqueloculina spp., Elphidium poeyanum, Hanzawaia spp., Discorbis williamsoni, D. floridana, Pyrgo sp., Cassidulina minuta, C. subglobosa,* and *Pararotalia cananeiaensis* reveal high dynamic currents and penetration of marine currents in the channel, with absence of eutrophication caused by TEBAR and urban sewage of São Sebastião and Ilhabela. Otherwise, the opportunist tolerant

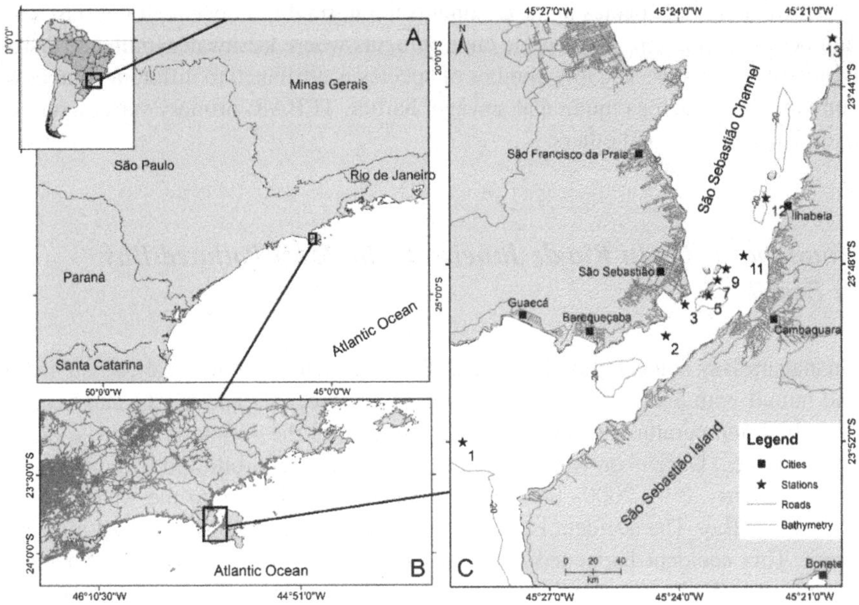


**Fig. 1.9** Study area and sampling stations for monitoring

group *Ammonia* spp., *Buliminella elegantissima, Bulimina marginata, Bolivina striatula,* and *Fursenkoina pontoni* are proliferating where high organic matter indicates terrestrial contribution.

The modern sedimentary processes are directly related to hydrodynamic circulation and bottom topography. The wind-generated currents are the most effective hydrodynamic phenomena, responsible for the bottom sedimentary processes, with currents greater than 1.0 $ms^{-1}$ velocities, and with the predominant northeast direction of current flux. The São Sebastião Channel presents variations in the patterns of diversity, dominance and evenness, relating to spring-summer and autumn-winter variations. These are characteristic of subtropical environments and related to channel oceanography. Additionally, in spring and summer, the South Atlantic Central Water mass, which carries high salinity and low temperature, enters through the southern entrance. Therefore the waters in the channel create microenvironments along the water column, and are responsible for the high phytoplankton diversity within the Channel.

We conclude that the foraminiferal associations in the Channel are not totally reducing. In addition, the large number of living organisms, young and adult, reveal that water circulation and dynamics favor the establishment and development of Forams, explaining the absence of eutrophication caused by TEBAR and the urban sewage of São Sebastião and Ilha Bela. However, the low number of species and the opportunistic species resistant to several types of pollution which proliferate in places where organic matter contents are high, indicates terrestrial contribution and anaerobic environments in the central regions of the Channel.

The diversity of species in the channel is not normal for inner shelf regions. This number is comparable to estuarine environments where freshwater limits the distribution of organisms. The low number of species is attributed to different sources of contamination such as municipal sewage, harbor, TEBAR, primary surface runoff, hydrocarbon spills and others.

## *Guanabara Bay in Rio de Janeiro Is the Most Polluted Bay in Brazil*

Guanabara Bay is located in the state of Rio de Janeiro. The tropical climate is hot and humid with average temperatures varying between 22 and 26 °C, and average daily high temperatures in summer are 30–32 °C. In the 4 months of the so-called high summer (December–March), "the very hot days" are followed by heavy rains.

On January 18th, 2000, 1292 $m^3$ of fuel oil leaked out into the waters of the Guanabara Bay. The accident is one of many oil spills occurring in Brazil in recent years. This accident happened north of Governador island, and affected a range greater than 50 $km^2$, reaching part of the wetland protected area of Guapimirim mangrove, and beaches along Guanabara Bay, causing death of fish and shellfish from contamination and lack of oxygen. Also causing serious social and economic concerns with severe damage to fishing and tourism. These costs that have not been compensated to local communities, which draw their livelihood directly or indirectly from the Bay (Fig. 1.10).

Twenty-six sediment samples collected in July 2000 and February 2001 were analyzed for pH, dissolved oxygen, and foraminiferal distribution. Foraminiferal species correlated with chemical effects related to the oil spill from the pipeline.

**Fig. 1.10** Guanabara Bay views in the state of Rio de Janeiro. A sampling map with an oil leak patch, and sediment contaminated with oil

Oil spilled into saline waters alters the oil's chemical composition allowing the breakdown of components. The oil spill releases $H_2S$ (hydrogen sulfide), ammonia and methane, causing the decrease of pH in sediment, and decrease of oxygenation in water. These changes interfere with the natural environment of the affected region, making it unstable.

In order to assess the environmental quality of the region and the damage to the marine ecosystem that occurred due to the oil leak, sediment and water samples in two seasons, winter 2000 (6 months after the accident) and summer 2001 (1 year after the accident) were collected. We used this material to evaluate the response of foraminifera and chemical data (pH and oxygen) in the face of this and other anthropogenic disasters.

The increase in sediment deposit in Guanabara Bay is an overlapping geological and natural siltation. Bottom sediments are dominated by silted sand and mud. The straightening and channeling of rivers and the destruction of mangroves also increases the volume of sand loads. Grain size distribution reflects the tidal current energy near the bottom, which is directly influenced by bottom morphology and the Guanabara Bay shoreline contour. The large amounts of organic mud is related to the eutrophication of the Bay's sewage and garbage, especially occurring at the entrance of the Bay.

The circulation and salinity in this bay are subject to tidal movements; wind and river discharge, and fits the model as a salt wedge estuary, where bottom waters are colder and higher in salinity. The highest salinities occur in the entrance of the Bay, and less saline waters occur near the mouths of rivers and wetlands, in the northern portion of the Bay, reaching salinity values below 8. In the central and south portions of the Bay, salinity ranges between 30 and 34.

Guanabara Bay receives huge amounts of mostly untreated industrial and municipal waste daily. The city of Rio de Janeiro Dumps sewage mainly into the waters of the S. J. Meriti River, located to the west of Governador Island. More than 6000 industries pour filth into the Bay's water. These include 2 commercial ports, 16 marine terminals, 2 naval bases, a shipyard, a large number of ferryboats, fishing boats and yachts, and the largest oil refinery in Brazil, Duque de Caxias (REDUC) dumps more than 1.4 tons of oil and industrial waste daily! And yet the crude oil leak shows ecological deleterious changes in Guanabara Bay!!

## *Ecological Accident: REDUC's Yuck!*

The low pH from the oil spill caused an increase of *Elphidium* spp. in winter (6 months after the accident in January 2000). This relative increase is mainly due to the absence of other species with thinner shells that were quickly dissolved by the toxic effects from the oil leak. Directly next to the REDUC oil refinery, the acidity of the sediment shows that the dissolution of carbonates in the region created complete sterility 1 year after the oil spill incident (Fig. 1.11).

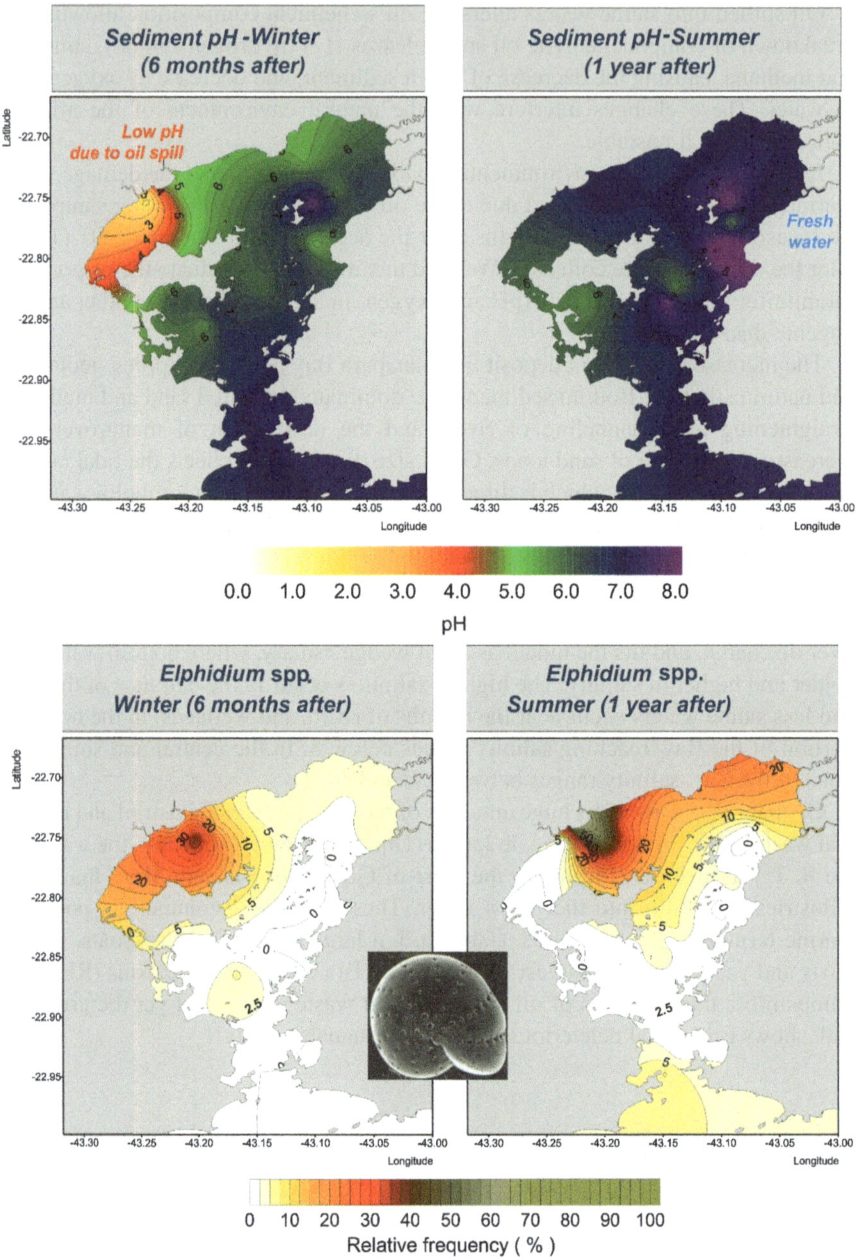


**Fig. 1.11** Sediment pH and relative distribution of *Elphidium* spp. in winter and summer

*Buliminella elegantissima* was found in the environmental preservation area, APA Guapimirim, in winter, although not close to REDUC, indicating that this species does not tolerate acidic pH environments, given the fragility of its thin shell. The damage assessment 6 months and 1 year after the accident was alarming since mortality and other symptoms in the population was still evident after 1 year (Fig. 1.12).

## *Domestic Sewage*

The oil spill reached an already stressed and altered ecosystem, Guanabara Bay is highly unregulated and overly urbanized. The dumping of sewage from the city of Rio de Janeiro takes place mainly in the waters of the Rio S. J. Meriti located west of the Governador Island. Our results show that south of Governador Island is seriously compromised because low efficiency of water renewal is unable to export the pollutants from the São J. do Meriti River.

In the northwest and central parts of the bay, low dissolved oxygen values are related to waste discharges, showing also dominance of *Buliminella elegantissima*. In winter, near Iguaçu River, sediment pH is low and highly acidic. The acidity is responsible for foraminiferal low diversity and limits the occurrence of species with fragile shells. On the other hand, thicker tests (shells) of *Elphidium* spp. were last to

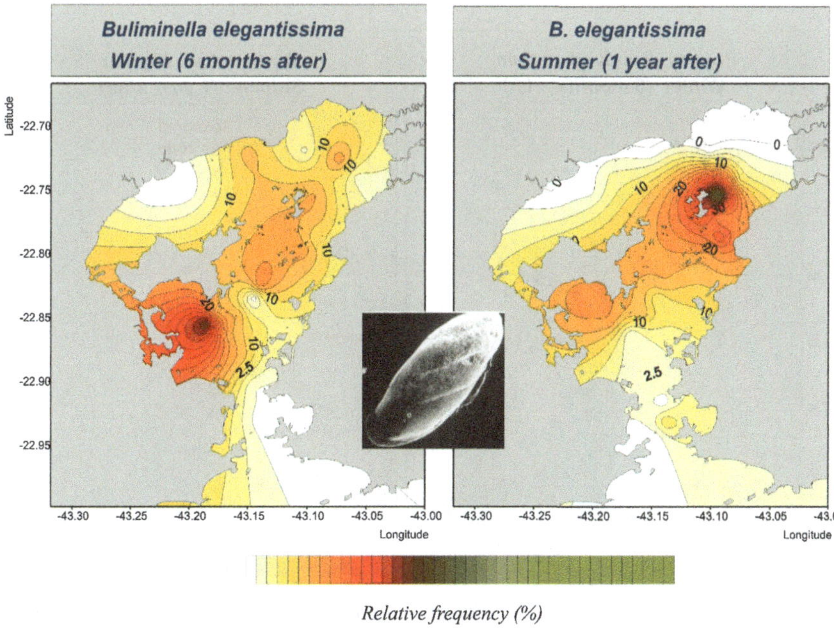


**Fig. 1.12** Relative distribution of *Buliminella elegantissima* in winter and summer

disappear, therefore increasing its relative dominance in winter. In summer, this became a barren zone and the dissolution of tests culminated with the total disappearance of these microorganisms 1 year after the accident.

The west margin and inner parts of the bay show high concentrations of nutrients (nitrogen, phosphate and sulfur) due to enrichment from domestic discharge and low renewal efficiency in the waters near Governador and Fundão islands. These locations are seriously impacted because the São J. do Meriti River influences the environments there, and north of the bay. This happens because west of the Governador Island, which would be the "natural" way of water flow, water is stagnated and closed due to silting and garbage littering in the region. Consequently, water passage tends to occur to the north, preventing water renewal.

Most of the bays are characterized by high concentration of oxygen in the sediment, which enables the maintenance of high biological activity. However, large amounts of organic material waste (sewage) can create areas where oxygen consumption exceeds supply, especially during summer months, making the environment anoxic. This is precisely what was observed in the south of the Governador Island. The central region of the Bay has the lowest bottom concentration of oxygen. This fact is also related to the increase in temperature this time of year and the phytoplankton "bloom", which results in areas of greater eutrophication and subsequent hypoxia (Fig. 1.13).

The seasonal study of Guanabara Bay shows periods of hypoxia similar to those observed in coastal areas like Chesapeake Bay, Gulf of Mexico, and the Continental Shelf of Louisiana. Our study clearly reflects loss of biological communities in rela-

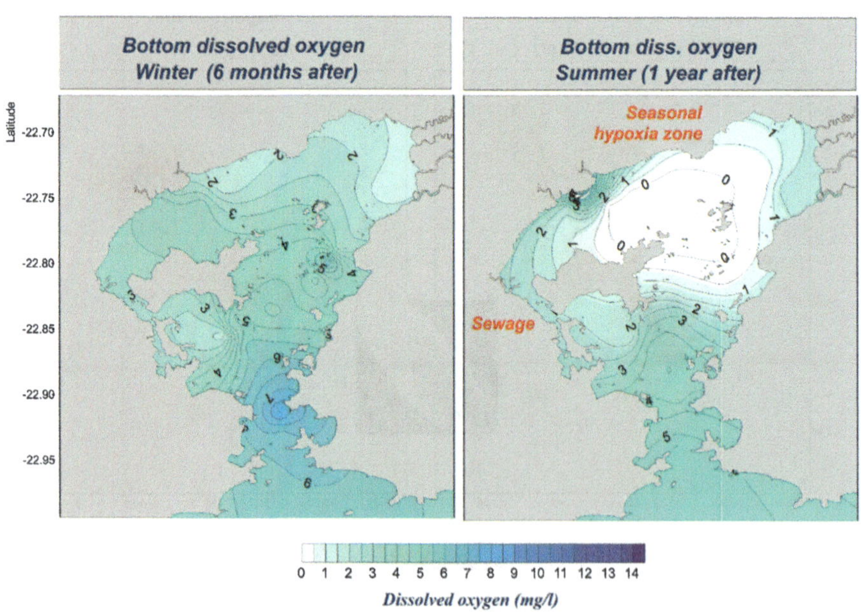


**Fig. 1.13** Bottom dissolved oxygen in winter and summer

tion to the rest of the Bay. South of Governador Island requires further attention since these regions experience higher quantities of organic dumping from the S. J. do Meriti River. Populations of bacteria and virus exist all over; and their development seem to be related to the amount of silt and clay in the sediment. The accumulation of silt and clay in Guanabara Bay is linked to both a "natural" source, influenced by the river run off, and anthropogenic sources, i.e., sewage outfall. Organic pollution is mainly associated with regular renewal of the water from each tide, favoring the establishment of opportunistic species.

*Buliminella elegantissima* is more abundant in the anoxic regions south of Governador Island and central bay. This finding reinforces that this species is typical of organic matter enriched sediments, with low oxygen, occurring in sediments contaminated by domestic sewage. In addition, this excessive nutrient enrichment favors the proliferation of opportunistic foraminiferal species, increasing local dominance with high growth rates.

Environments with high organic matter concentration are directly related to higher concentrations of fine sediment. This type of sediment is characteristic of low energy environments, which facilitate the deposition of organic matter. Moreover, in these regions, the oxygen concentration is very low. *B. elegantissima* presented its highest relative frequency exactly where hypoxia is enhanced in the region south of Governador Island in winter, and in the central region near the anoxic zone in summer. Anoxic zones are usually located where the increase of bacteria from extreme concentrations of pollutants decreases oxygen concentration.

The hypoxic region of REDUC has greater limitation in productivity although its oxygen concentration is similar to the APA. The dominance values are high both in the region closer to REDUC and in the central part where anoxic zones reflect the establishment of a few species adapted to those polluted and contaminated conditions.

Guanabara Bay is highly impacted, dominated by few a species with a low diversity of fauna. The number of species per sample increases when environmental conditions are more typically saline, but if compared to other less saline lagoons and estuaries in Brazil, Guanabara Bay should have more living foraminiferal species.

## *Todos os Santos Bay (Bahia State)*

The environmental quality of Landulpho Alves Oil Refinery (RLAM/PETROBRAS) situated on Todos os Santos Bay (TSB) in Bahia was quantitatively and qualitatively assessed in different environments for 2 consecutive years (Sub-tidal area, and Caboto and Camamu controls/subtidal, July—December 2003, July 2004—January 2005) (Fig. 1.14).

Marine and mixohaline species (such as *Bolivina* spp., *Bolivina pulchella, Pseudononion atlanticum, Fursenkoina pontoni, Buliminella elegantissima, Bolivina striatula, Bulimina marginata, Quinqueloculina spp., Ammonia spp.,*

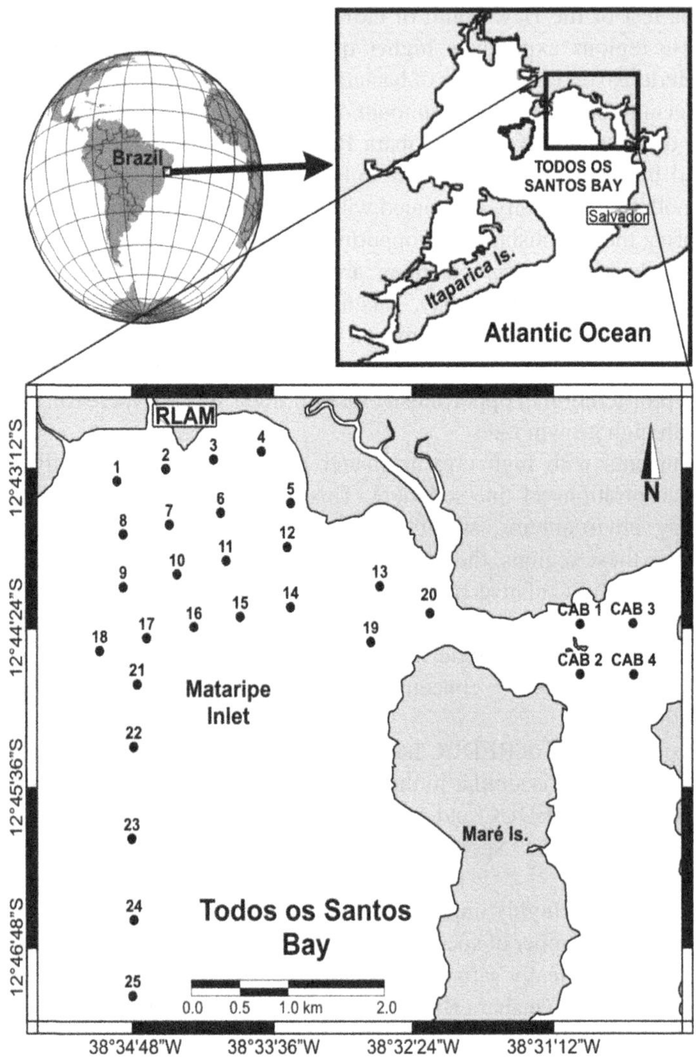


**Fig. 1.14** Sampling map and studied area

*Elphidium* spp.) with low influence from continental runoff, are tolerant to high levels of accumulated organic matter, in the southern outermost part of the Bay. These species show tolerance to sediment acidification, indicating that areas closer to the ocean, which should display higher species diversity, decrease instead.

The type of fauna found in Todos os Santos Bay (TSB) indicates that marine waters dominate the environment. On the other hand, the low diversity and strong dominance of a few species is characteristic of poorly preserved estuarine environments where fresh water limits organism distribution.

*Buliminella elegantissima* is the species typical of environments rich in organic matter, which is mainly retained in muddy sediments, found practically in all regions of TSB. Environments with high organic contents have a direct relation to high mud concentrations (Tyson 1995), since in this type of sediment energy is quite low, contributing to the processes of organic matter deposition. In addition, fine sediments have a tendency to retain heavy metals and other pollutants (Kjerfve et al. 1997).

The number of species per sample increases when the environmental features are more typically marine (Phleger 1970). Thus, it was to be expected that TSB presented a higher diversity per sample level than more restricted estuarine environments.

Although TSB is under the influence of continental runoff in the summer and winter, the foraminiferal assemblages indicate that this influence is quite small. Comparison of the present study with the data obtained by Eichler-Coelho (1996) in the Cananéia-Iguape estuarine system (São Paulo State) and the data by Debenay et al. (1997) in Conceição Lagoon (Santa Catarina State) shows that the number of species found is similar to what we have found in the present study. This fact demonstrates that the pollution effects superimpose themselves onto the natural environmental factors, limiting the establishment of non-opportunistic species and facilitating the development of the opportunistic ones.

Species that are abundant in polluted areas are obviously tolerant to the pollutants to which they are subjected, whereas other species can manifest their sensitivity to the same pollutants by their absence (Yanko et al. 1994).

The accumulation of pollutants in coastal areas does not rely entirely on the supply of these materials by rivers, but also from the chemical interaction between those elements and the sediment constituents, and are therefore reflected by the carbon and fine sediment contents which more easily adsorb these types of particles. Such environments present high food concentration for several opportunistic species, a fact revealed in the present study by the occurrence of *Bolivina* spp., *Bolivina pulchella, Pseudononion atlanticum, Fursenkoina pontoni, Buliminella elegantissima, Bolivina striatula, Bulimina marginata, Quinqueloculina* spp., *Ammonia* spp., and *Elphidium* spp. The foraminiferal genera *Ammonia* and *Elphidium* are indicators of pollution levels where *Ammonia* is more tolerant of polluted environments, and *Elphidium* is less tolerant. *Ammonia tepida* is also resistant to low concentration of dissolved oxygen (around 3 mg/l) and high sulfur values characteristic of toxic sediments.

Compared to the samples collected in the RLAM subtidal area, the samples collected in the Caboto and Camamu control regions presented few individuals, low species numbers and the smallest diversities, suggesting that the control areas are not environments like those in the RLAM area, and therefore are not wholly comparable. The diversity variation at the stations may be linked to the amount of organic matter available and to the capacity to retain this food source used by the foraminifers. Sediments with high predominance of silt and clay contain great amounts of retained organic matter making the environment suitable for the establishment of different species.

Analysis of relative abundances suggests that the species *Gaudryina exillis*, *Textularia earlandi*, *Arenoparrella mexicana*, *Haplophragmoides wilberti*, *Trochammina* sp., *Ammotium* spp., *Ammoastuta inepta*, and *Miliammina fusca*, which occur in places more subjected to continental runoff, were grouped because its similarity of occurrence.

The marine species *Globocassidulina subglobosa* and *Uvigerina* spp. are species rarely found in the study region, since they are usually associated to colder water masses. In the south region of the RLAM subtidal area, the penetration of the marine current was evidenced by the presence of *Discorbis* spp. in that region. Evenness is distributed along a gradient, where the lowest environmental stability values occur close to the RLAM and the Mataripe River continental runoff, indicating less environmental stability in those regions, and increase of stability toward the region more subjected to ocean influence to the south. As environmental stress increases, species diversity falls, resulting in an increase of dominance (Odum 1988). The north-to-south diversity increase probably reflects local hydro dynamism, where the saline waters are responsible for water renovation in the southern portion of TSB. The data obtained in the present study reveal that the northern sector, including the regions close to the RLAM, are low-circulation zones, and are thus less subject to water renovation. The analysis carried out over 2 consecutive years shows a relative environmental stability in terms of foraminifera, and do not have an apparent relationship to eventual inter-annual changes.

## *Sedimentary and Coral Reef Areas in the Shelf of Rio Grande do Norte State*

Sometimes called "rainforests of the sea", coral reefs are living polyps held together by carbonate exoskeletons and are some of earth's most diverse ecosystems. Reefs occupy only 0.1% of the world's ocean area, while providing a home for an astonishing 25% of all marine species. East and West of Natal, the state capitol of Rio Grande do Norte, sedimentary and coral reef fields of Pirangi, Maracajaú and Açu occur inshore and in inner-shelf environments (Fig. 1.15).

The Açu reefal area is a mesophotic-submerged reef, newly discovered in 2015 by Gomes and collaborators in the oriental shelf area at 60–85 m depth. The small reefs of Pirangi are about 25 km south of Natal in <5 m of water. The Maracajaú is part the Sioba, Cação, Rio do Fogo, and Maracajaú coral reef chain, 60 km north of Natal, and are located 5–7 km from the coast. This reef lies ~5 km offshore at depths of 3–5 m, and consists of knolls and pinnacles that reach up to 6 m in height, and may be partially exposed at low tide (Santos et al. 2007). After its discover, the Açu coral reef area was under investigation and the upwelling of cold waters phenomena associated to foraminiferal assemblages was documented by Eichler et al. (2019a, b).

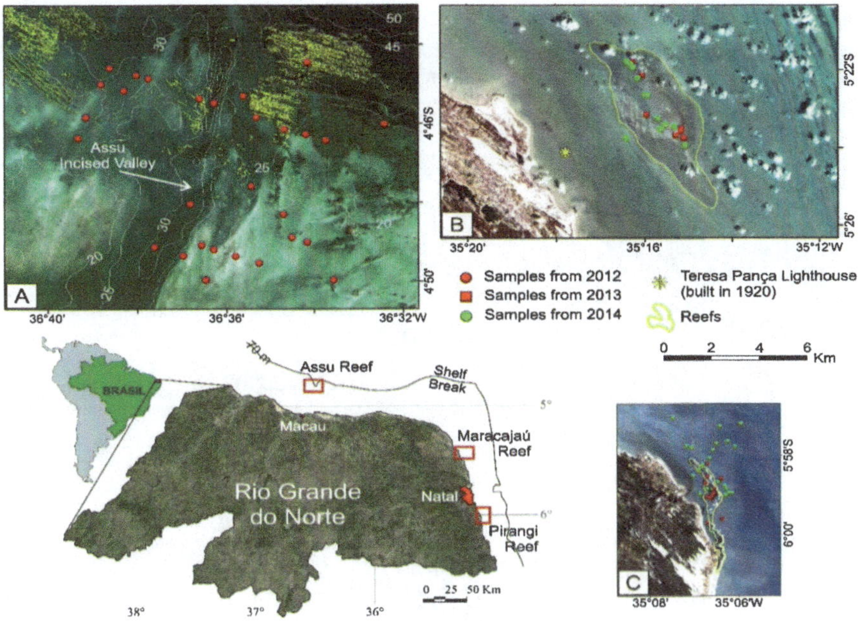


**Fig. 1.15** Sedimentary reef areas. (**a**) Acú. (**b**). Pirangi. (**c**) Maracajau

The shelf is narrow with less than 40 km wide, and shelf break is at ~70 m water depth (Vital et al. 2010a, b; Gomes et al. 2014). Subaqueous dunes and shallow isolated sand bodies are present in the inner shelf. Submerged beach rock marks the middle and outer shelves, and two incised valleys cut the shelf from the coast to the shelf break (Testa and Bosence 1999; Santos et al. 2007; Vital et al. 2008; Gomes et al. 2015, 2016).

Among the anthropogenic activities in the reefal Areas, are the "Marina Badauê Boat Tours" which take tourists to explore the Natural Pools of Pirangi, 800 m from the coast (Fig. 1.16a, b). They leave from Pirangi Beach three times a day year-round; the tour itinerary lasts an average of 2 h, with a stop for swimming and diving in the natural pools (Fig. 1.16c, d).

At low tide, reefs are exposed and small pools are created where small colorful fish can be observed without the use of a mask. The tour company promotes "walking on the reefs" and rents sandals for that purpose (Marina Badaue 2017). In addition to direct impact by tourists, the close proximity to the coast also exposes the Pirangi reefs to domestic and industrial waste from the Pirangi watershed.

Maracajaú Reef is considered one of the ten best places in Brazil for snorkeling and scuba diving. At low tide, the area transforms into natural pools of crystalline water with depths of 1–3 m beautifully rich in diverse marine flora and fauna.

Figure 1.17a, b show tourist boats at the Maracajau reefal area; the visitors are swimming and snorkeling rather than walking on the reef (Fig. 1.16d), in water depths greater than at Pirangi. The Maracajaú reefs belong to a conservation area

**Fig. 1.16** (**a**) General view of Pirangi reefal area, arrow points to four tourist boats. (**b**) Detailed view of the position of the boats on the reefs. (**c**) Many people in the water three times a day year-round. (**d**) Is it fun to walk on and damage the living reefs?

called the "Environmental Protection Area of Coral Reefs" (APARC), created in 2001. According to Araújo and Amaral (2016), the APARC and the adjacent continental shelf are currently experiencing an increase in environmental degradation due to overfishing, coastal development, and increasing tourism.

In Pirangi reefal area, main structures of support for living corals and other reef inhabitants occur as patches, constructed primarily of dead sponges and algae, instead of sedimentary rocks composed of coral exoskeletons typical of coral reefs in other parts of the world, such as the Great Barrier Reef, Australia (Maxwell 1968, 1969; Maxwell and Swinchatt 1970; Flood and Orme 1988). Maracajaú patch reefs occur on sand stones that formed as beachrock (Castro and Pires 2001; Martinez et al. 2012). Brazilian reefs are distinct from reefs of Caribbean, Indian, and Pacific oceans with regard to their faunal composition; there are fewer coral species, several of which are endemic (Leão et al. 2016). Fleshy macroalgae and turf algae also compete for space on the reefs.

High algal abundance (Francini-Filho et al. 2013) and low coral cover (Barradas et al. 2010; Medeiros 2009; Azevedo et al. 2011; Krajewski and Floeter 2011; Martinez et al. 2012) are common substrata, as we observed at Pirangi. Although the dominance of algae is generally considered a negative health indicator for coral reefs (Venera-Ponton et al. 2011), this statement is not necessarily true for all reef ecosystems, especially those with relatively low coral diversity. In some Hawaiian

**Fig. 1.17** (**a**) General view of Maracajaú reefal area. (**b**) Presence of many tourist boats. (**c**) Detailed view of visitors swimming instead of walking on the reef. (**d**) Relative depth showing a swimmer being very conscious of protecting the reefal area

reefs, for example, there is high percentage of macroalgae cover; however, all essential ecological processes remain intact, and algae are considered important contributors to the general health of these reefs (Vroom et al. 2006; Vroom and Braun 2010), similar to our Pirangi reefal site. Sponges represent another major component of substrate cover.

The three-dimensional sponge structures offer complex spaces that increase the variety of microhabitats and elevates the diversity of local fish and invertebrates (Buhl-Mortensen and Mortensen 2004; Mortensen and Buhl-Mortensen 2005; Souza et al. 2015).

The distributions and abundances of the tests of symbiont-bearing benthic foraminifera can offer valuable clues to the health of a coral-reef ecosystem (e.g., Hallock et al. 2003; Barbosa et al. 2009, 2012). Variations in the total abundances of the two most common species, *Amphistegina gibbosa* and *Amphisorus hemprichii*, show that overall, in the sediments of the Rio Grande do Norte reefs, *Amphistegina* tests were far more abundant than *Amphisorus* tests. Amphisorus are more abundant were people walk upon the reefs than Amphistegina. It may be related to its flat morphology, capable of living under things which tolerate heavy weights (Fig. 1.18).

However, because we were dealing only with the presence of tests in sediments, bottom water hydrodynamics must also be considered when assessing the distributions of these tests. Sediment data from the parts of Pirangi and Maracajaú reefs

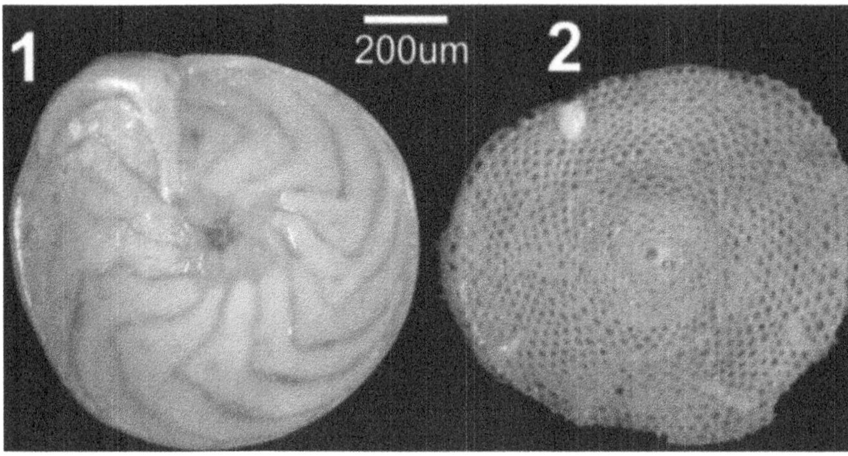


**Fig. 1.18** *Amphistegina gibbosa* and *Amphisorus hemprichii*

affected most by tourist's anthropogenic disturbance activities revealed that *Amphistegina* tests were less abundant here. *Amphistegina* was not abundant at any of the stations sampled in the tourist area at Pirangi in 2013, but had low numbers at four stations within the tourist area. In 2014, *Amphistegina* was abundant at ten of the 11 tourist-frequented areas in Pirangi. At Maracajaú in 2013, the highest numbers of *Amphistegina* tests were found in the reef-margin area, and were absent where reefs were trodden upon. *Amphisorus* tests were found even where people often walk, mainly on soft corals and seagrass beds. *Amphisorus* is typical seagrass-associated species (Reich et al. 2015; Hohenegger 1994). They found them living in abundance on macroalgae on reefs in the northwestern Pacific, and on vegetation free substrates were noted the same in Pirangi, Maracajau and Açu coral reef areas. *Amphistegina* spp. are usually found associated with reefal or firm phytal substrates, and *A. gibbosa* generally does not live abundantly in very shallow waters (Hallock 1999; Baker et al. 2009). Tests of the two species of foraminifers assessed were particularly depauperate in the Pirangi samples collected in 2013 in 12 samples out of 30, and 4 samples out of 25 in 2014. At Maracajaú in 2013, tourists were diving close to the impacted area which was not sampled in 2014. In the more extensive patch reef system of Maracajaú, the community appeared healthier, except where tourist activities impact had physically damaged the reef substrate.

Coral death from anthropogenic disturbance is a particular concern for reefs that are walked upon, or otherwise heavily visited by tourists, but reefs nearby rivers are also at risk. Agricultural land use has increased the amount of sediment, nitrogen, and phosphorus accumulating in coastal-marine environments and high turbidity blocks light penetration, which is indispensable for symbiont-bearing foraminifera (Fujita 2004).

Among the studied reefs, only those at Pirangi are very close to a major city (Natal, the state capital). The shallowest part of the Pirangi reefal area, closest to Pium River, is expected to suffer negative effects. However, *Amphistegina* was common and abundant at the closest stations sampled in 2013. The Maracajaú reefal area is farther away from the coast, compared to Pirangi, and the foraminiferal tests appeared better preserved, although the relative distributions of *Amphistegina* and *Amphisorus*, and the impact of trampling are similar.

Many foraminiferal shells in our samples from the Pirangi reefs were broken, corroded, unusual brownish and yellowish discoloration, with few living individuals, indicating reworking of the sediment. Araújo and De Jesus Machado (2008) and Oliveira-Silva et al. (2012) also found broken and corroded tests on the Abrolhos offshore carbonate shelf reefs, near Bahia state. *Quinqueloculina* had discolored and eroded shells and were reported by Oliveira-Silva et al. (2005) at sites hydro dynamically influenced by the high volumes of water carried by the Doce River onto a narrow shelf. It is possible that Pium River has a similar reworking effect at Pirangi. The eroded and broken tests of specimens suggest that some components of the foraminiferal assemblage in seafloor sediments off the Rio Grande do Norte coast may be relict. In particular, brown or black specimens of *Quinqueloculina lamarckiana* dominant porcelaneous species (without algal symbionts), are widespread and indicate that erosion may be taking place in the inner shelf of the area, reworking old sediments.

# Chapter 2
# Deep in Mud: Benthic Foraminifera Interface Marine Sediment-Water—Sampling, Handling, and Analyzing

**Abstract** Sampling shallow or deep environments requires different techniques due to particular and irregular morphology in the marine sediments worldwide. Independent of the way sampling is performed, systematic methodology for handling and analyzing are similar. Measurements of surface and bottom water properties, generally done at the time of collection, help us identify bottom water mass and its link to specific living benthic assemblages. When analyzing marine sediment samples, a qualitative approach where main indicator species are selected is based on the relationships between their specific distribution patterns and the bottom hydrographic data (salinity, temperature, dissolved oxygen, suspended matter, pH, silica, nitrite, and nitrate), and geological data (grain sizes, carbonates, and organic matter) and other available parameters are a good choice when detecting patterns. Dead shells are observed in relation to the state of test conservation due to acidity and transportation, and occurrence and extinction of specific species delimits stages, epochs, and periods of time can be quite interesting. Quantitative analysis based on counts of living specimens can be descriptive or analytical in recent and in the past millions of years. Numbers of individuals are counted and the individual absolute and relative abundance (dead/unstained and alive/stained) are calculated. Maps of spatial distribution of relative abundance of species are important to illustrate and understand tendencies since each datum relates to its geographic coordinates. Univariate and multivariate analysis, ecological indexes like diversity, dominance, evenness, hypoxia (AE index), FORAM Index (FI), and the paleo temperature ("Warm" species index) are applied to foraminiferal and environmental data to further unravel particular situations, and are effective to assess changes in community structure helping us to find the problem and propose a better solution for the best quality of marine sediment through maps. Mapping the placement of samples reflects the similarity of their biological communities, environmental patterns, and indicators of biodiversity and stability, rather than their simple geographical location. Recently the association between 'warm' species index, organic carbon flux, and modern bottom water temperatures has started to show excellent results on depth and differentiation of relatively warm groups (>3.5 °C) with higher carbon flux (>3.5 g C $m^{-2}$ $year^{-1}$) versus a group that reflects the colder, deeper waters with lower carbon flux in a deep water section of the Western Pacific Warm Pool, the

 
P. P. Beck Eichler, C. P. Barker, *Benthic Foraminiferal Ecology*,
https://doi.org/10.1007/978-3-030-61463-8_2

warmest area of the Pacific. Relatively high abundances of an indicator species (*Uvigerina prosbocidae*) of paleo productivity from organic matter input associated with down-slope movement in a tectonically unstable region shows co-variates with 'warm' species groups, providing evidence for the use of this species as a tracer for carbon flux. Marine stable isotopes (δ $^{18}O$ and δ $^{13}C$) also can detect salinity and temperature patterns near freshwater sources or near the ocean. Historical changes in the mixing between fresh versus marine waters of bays and estuaries should follow spatial sense, and they can be determined by analyzing faunal sequences and geochemical signatures in cores of its substratum. By applying both qualitative and quantitative approach we have a better knowledge of the environment and the details are not lost.

**Keywords** Sampling · Statistical analyses · Biodiversity · Stability · Accurate proxy · Marine environments · Ecological indexes · Forecasting

## Introduction

This chapter covers methods and techniques of sampling, handling, and analyzing benthic foraminifera from the interface of marine sediment-water. It explains how to sample in shallow and deep marine environments, how to process samples in the laboratory and analyze them statistically with descriptive and analytical methods. This includes how to wash shells from silt and clay in water using different sized sieves, separating shells from sediment grains, and how to identify and count species and genera. Apart from sampling and processing, we introduce useful common statistical methods to correlate biological distributional patterns and environmental data that unravels interesting multi-interdisciplinary findings in ecology, biology, geology, oceanography, geochemical, climatology, and paleontology, which permits us to accurately forecast biodiversity and stability patterns in the marine environment.

Here we will also show how the abundances and indexes of foraminifera are systematically prepared and studied as useful and meaningful statistically accurate proxy that correlates with different abiotic parameters available for specific problems to evaluate, and solutions to propose. We demonstrate how to use the simplest and easiest techniques that provide the desired results for each particular environment.

As an aside, we love everything about Forams, including processes and preparations for their research and study, and we hope to impart with the reader the joy they bring to us. Moreover, as a reminder, shells and 'tests' are interchangeable descriptions.

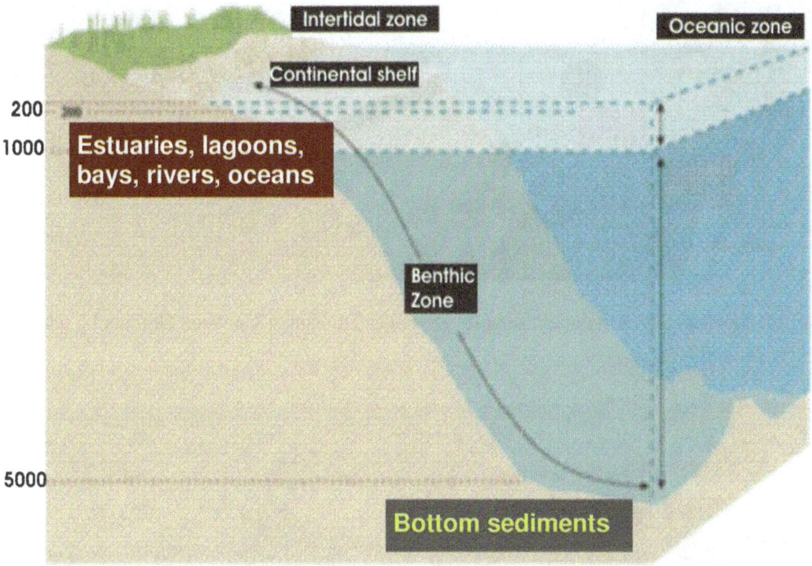


**Fig. 2.1** Benthic Foraminifera lives in thes bottom sediments and Planktonic Foraminifera lives in the water column of all coastal zones.

## *Where Do Forams Live? Our First Figure Illustrates That!! (Fig. 2.1)*

# Material and Methods

## *Sampling Shallow Environments (Inner Shelf and Reef Areas)*

Small vessels are required to sample marine surface sediment samples collected using a Van Veen Grab sampler (Fig. 2.2) in water depths of 200 m maximum, using a common rope or a winch for deeper areas.

Scuba divers collect marine sediment samples where a Van Veen grabber cannot reach due to irregular morphology around reefs. Underwater video and photographic surveys are also taken (Fig. 2.3).

Samples either collected with Van Veen grabs, from divers, or with core equipment are sub-sampled with spatulas. About 20–50 $cm^3$ of sediment is removed, preferentially from the top layer (1 cm). Figure 2.4 shows sampling process. After it is scraped off, the uppermost sediment sample layer is kept in ethanol with Rose Bengal for staining the live specimens. Staining is important for visual acuity of living Forams.

**Fig. 2.2** Marine surface sediment samples are collected using a Van Veen Grabber by hand using a rope and a winch

**Fig. 2.3** Scuba divers collect sediment samples where Van Veen or core equipment cannot reach due to irregular morphology of reefs. Underwater video and photographic surveys are taken

## *Sampling Deep Marine Environments*

Cores are collected using large vessels fitted with drilling booms, which are able to collect hundreds of meters of marine sediment at great depths. Figure 2.5 shows the JOIDES Resolution (JR) research vessel of International Ocean discovery Program (IODP), drilling and coring equipment working deep underwater, and marine sediment cores exposed to be described and sampled.

Marine sediment core samples tagged and bagged for study are described in Fig. 2.6 showing that many people are involved in the task of sampling deep cores.

Measurements of surface and bottom water properties are generally also done at the time of collection (Fig. 2.7). These measurements are taken into account when studying water mass parameters of salinity, temperature, amount of dissolved oxygen, and so on; they help us identify bottom water mass and its link to specific living benthic assemblages.

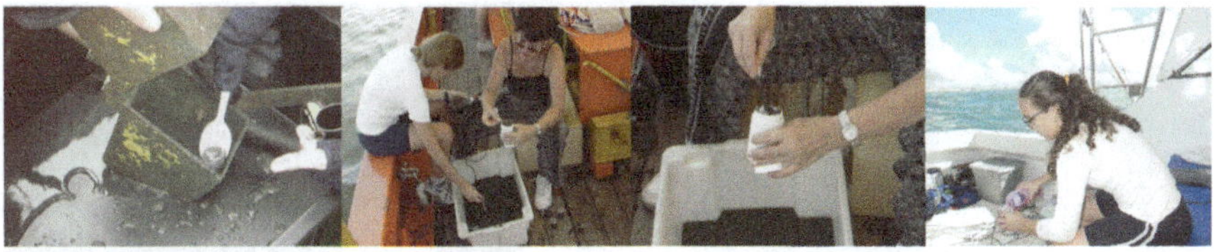

**Fig. 2.4** After collection, the uppermost sediment sample layer (about 1 cm) is scraped off and kept in ethanol. A solution of Rose Bengal in ethanol is used for staining live specimens classifying them like (dead/unstained and alive/stained)

**Fig. 2.6** Below deck, where marine sediment core samples are tagged and bagged for study

**Fig. 2.5** Large vessel (JR, IODP) drilling for deep water marine sediment core samples. Cores being collected. Different sediment layers to be described for geological patterns and facies

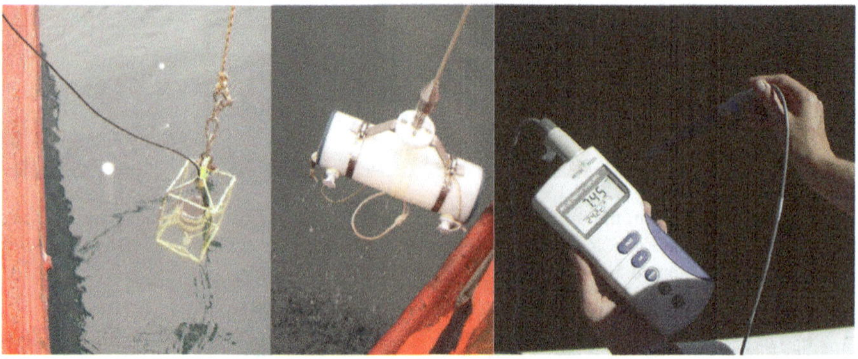

**Fig. 2.7** Measurements of abiotic parameters of water surface

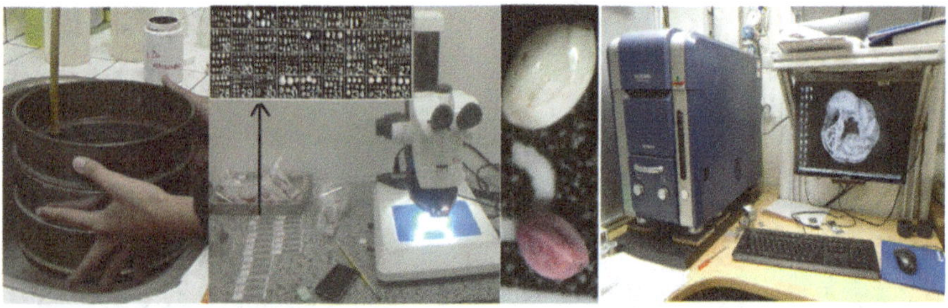

**Fig. 2.8** Sediment washed through 0.500, 0125 and 0.062-mm sieves for removal of silt and clay. Foraminifera is brush-transferred to slides with a black background for later identification in a stereo microscope. Bengal Rose unstained and stained (living and dead) foraminifers' shells. Scanning microscopic electronic photographs of species

## *Handling Marine Sediments*

After staining for 48 h with Bengal Rose to identify living specimens from dead ones, a fixed volume of 50 $cm^3$ of sediment is washed through 0.500, 0125 and 0.062-mm sieves for removal of silt and clay, and then oven dried at 60 °C. Biological material is sorted, and foraminifera is transferred with the aid of a very fine brush to slides with a black background for later identification. Counting is followed by the common practice of using a 125 μm dry sieve, and if necessary, splitting samples with a microspliter until about 100 benthic foraminiferal tests can be counted in a given fraction. To avoid contamination of foraminifera between samples, sieves are thoroughly cleaned, placed into a sonicator for at least 15 min, and then carefully checked under a microscope. A stereo microscope is needed for species identification. Scanning microscopic electronic photos are then taken to illustrate and identify each species (Fig. 2.8).

## *Analyzing Marine Sediment Samples*

### Qualitative Analysis

The habitat preference of several foraminiferal indicator species is described as characteristic of determined type, a feature of sediment, or water that the creature was living in at the time it was collected. Main indicator species are selected based on the relationships between their specific distribution patterns and the bottom hydrographic data (salinity, temperature, dissolved oxygen, suspended matter, pH, silica, nitrite, and nitrate), and geological data (grain sizes, carbonates, and organic matter) and other available parameter). Figure 2.9 show how of abiotic factors affect ecological dynamics with its peculiar environment.

In the tests fragmentation science, called taphonomy, dead shells are observed in relation to the state of test conservation, due to acidity and transportation. The occurrence and extinction of specific species delimits stages, epochs, and time periods due to the ability to preserve past environments reflecting present and past climates.

Figure 2.10 shows examples of different species as indicators of a specific past environment. They are responsible for dating the age of rocks and sediments (stratigraphic correlation) to ~500 million years.

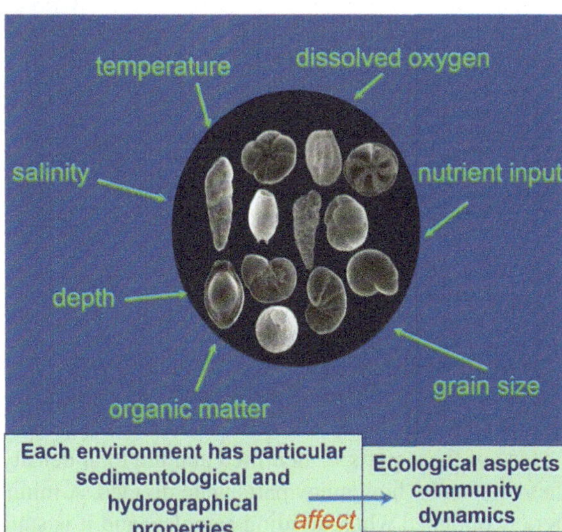


**Fig. 2.9** Some examples of abiotic factors affecting ecological dynamics

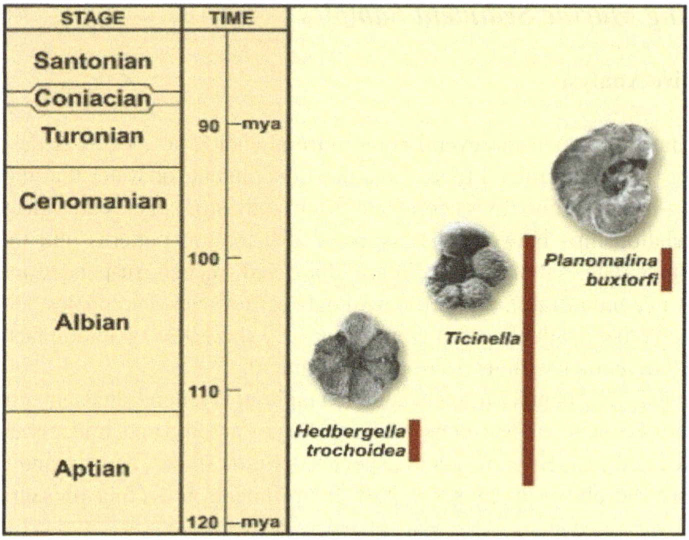


**Fig. 2.10** Occurrence and extinction of specific indicator species delimits stages, epochs, and time periods in the past

**Fig. 2.11** A microsplitter

## Quantitative Analysis

Primary dataset of quantitative analysis is based on counts of living specimens. If a sample has too many tests, the sediment content is divided into parts using a micro splitter (Fig. 2.11).

One of these parts is identified and the total density is estimated by multiplying these counts by how many parts were divided. A minimum of 100 tests of foraminifera is obtained when dividing samples, and it is statistically excellent to evaluate patterns worldwide. Scanning electron micrographs are taken to help with problematic identifications. Absolute and relative abundances are computed for each species. Numerical analyses are performed afterwards; and are described below.

## Descriptive Statistics

From the identification of Foram species found in study regions, numbers of individuals are counted and the individual absolute abundance (dead and alive) per sampled station is obtained. From the total absolute abundance, the species' relative abundance is calculated.

Maps are drawn up of the foraminiferal and abiotic parameters spatial distributions using systems of geographic referenced information, which relates each datum to its geographic coordinates (Fig. 2.12).

## Analytical Statistics

Univariate analysis is used to describe and summarize foraminiferal distribution evidencing central tendency (mean, mode and median) and species dispersion: range, variance, maximum, minimum, and standard deviation by looking at numbers of individuals of species. This data with ecological indexes like diversity (Shannon-Wiener, Newman 1995), dominance (Simpson), and evenness (Pielou) is effective when used together to assess changes in community structure (Zar 1984). It helps us to find the problem and observe the best quality of environment due to indicators of biodiversity and stability.

Multivariate analysis are applied to both environmental and foraminiferal data to further unravels data of the proposed environmental situation.

Environmental parameters of the area are then analyzed using Principal Components Analysis (PCA) to evidences groups of stations that are more similar due to the analyzed environmental characteristic, and an example of an output of

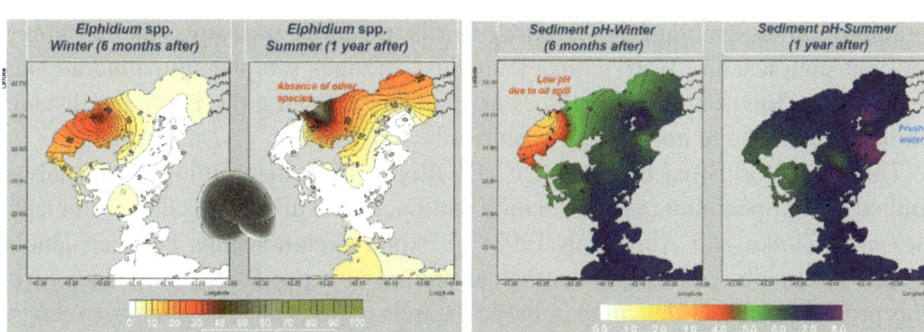


**Fig. 2.12** Distributional maps of foraminiferal spatial distribution and abiotic parameters

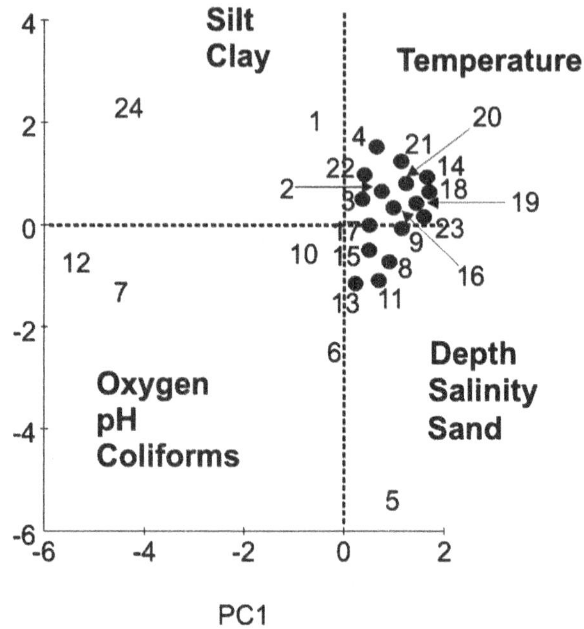


**Fig. 2.13** An example of Principal Components Analysis (PCA) and its groups formed in accordance to the abiotic variable

this analysis can be visualized on (Fig. 2.13). There are tables and results as other outputs but they are not to be mention here.

Non-Metric Multidimensional Scaling (MDS) on the other hand uses foraminiferal data methods to produce a "map" in which the placement of samples reflects the similarity of their biological communities and environmental patterns, rather than their simple geographical location. These descriptive ordination techniques provide a visual expression of complex data in two dimensions without assumptions of dimensionality inherent in other ordination techniques; sample separation in ordination space generally corresponds well to sample dissimilarity in ecological and environmental space (Fig. 2.14).

Biological matrices are constructed using the Bray-Curtis similarity measure on the log(x + 1) to normalize foraminiferal counts. Abundance data are calculated for each foraminiferal species and are subjected to Q-mode cluster analysis to define foraminiferal assemblages. The Bray-Curtis distance measures the proximity between samples, and Ward's linkage method arranges samples into a hierarchical dendrogram. Environmental matrices are constructed using the Euclidian distance similarity measure to normalize variables.

The BEST or Biota-Environment (BIOENV) are other multivariate analysis applied using Spearman rank correlation and Euclidean distance as the resemblance measure (Clarke and Ainsworth 1993) to explore relationships between abiotic parameters, main species and ecological indices at a $p < 0.05$ significance level. This procedure allows matching of biotic datasets to find the best match between the multivariate patterns of the assemblages.

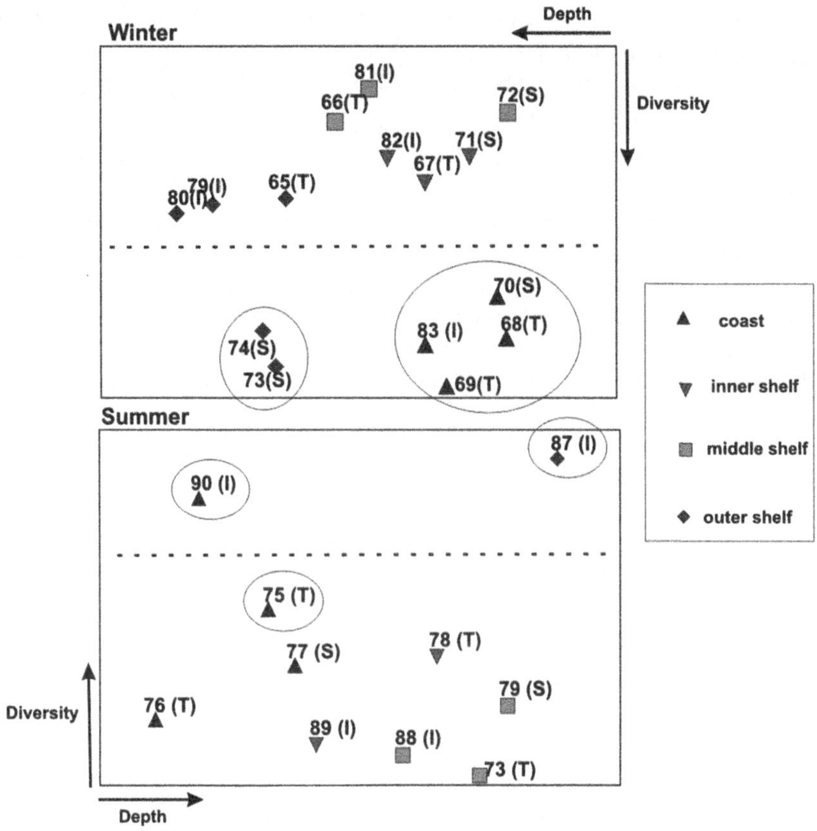


**Fig. 2.14** Non-Metric Multidimensional Scaling (MDS) analyses of foraminiferal data and group stations in relation to depth

The numerical procedures of Diversity, Dominance, Evenness, PCA, Cluster, MDS, and BIOENV can be carried out using the University of Plymouth computer program PRIMER v6 (Plymouth Routines in Multivariate Ecological Research PRIMER 6 Ltd., Plymouth), which is described by Clarke (1993), Clarke and Ainsworth (1993), and Clarke and Warwick (1994a, b), or other similar available computational routine.

## Foraminiferal Indexes

- AE Index

  To assess the degree of environmental pollution, the hypoxia index (A-E index) described by Sen Gupta et al. (1996) is a good choice. This index is currently utilized to determine the degree of sediment hypoxia according to the absolute abundance of two genera common in coastal environments (*Elphidium* and

*Ammonia;* Fig. 2.14). Species of these genera are tolerant to hypoxia, although species of *Elphidium* are less resistant than those of *Ammonia.*

The A-E index is calculated by the formula:

- $$\frac{A.B.Ammonia}{A.B.Ammonia + A.B.Elphidium} \times 100$$
- Where: A.B. = absolute abundance (number of individuals/50 cm$^3$).

Values of the A-E index are plotted in scatter graphs that correlates the percentage of environmental abiotic to this index. These indexes can be applied in recent and paleo data. An example from the Long Island Sound: The last 40 years (Thomas et al. 2000) in the Long Island coastal zone (New York, USA) using the AE index and were able to track oxygenation, influx of 'fertilizer' and stratification and stagnation of water and the episode of anoxia (no oxygen) in 1971 (Fig. 2.15).

Another study using AE index were performed in Osaka Bay (Japan), where the eutrophication on shallow marine benthic over the last 150 years was assessed by Tsujimoto et al. (2006a, b) that found out four stages of pollution and verified Shannon index (diversity) and evenness also show agreement with environmental interpretations and AE index (Fig. 2.16).

- A downcore study of stable isotopes in foraminiferal tests test the following in coastal areas:
- (a) possible link between hypoxic events (algae proliferation, and the fish kills) and recent anthropogenic induced changes;
- (b) when such events started in time;
- (c) if these hypoxic events are merely natural cyclic events or human induced.
- FORAM index (FI)

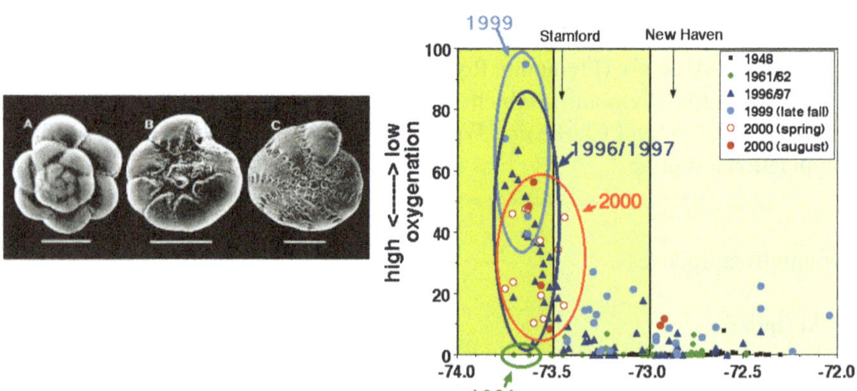


**Fig. 2.15** *Ammonia* (*A* dorsal view, *B* ventral view) and *Elphidium* genera (*C*). An example of the use of the AE index showing decrease in oxygenation from 1961 to 1999

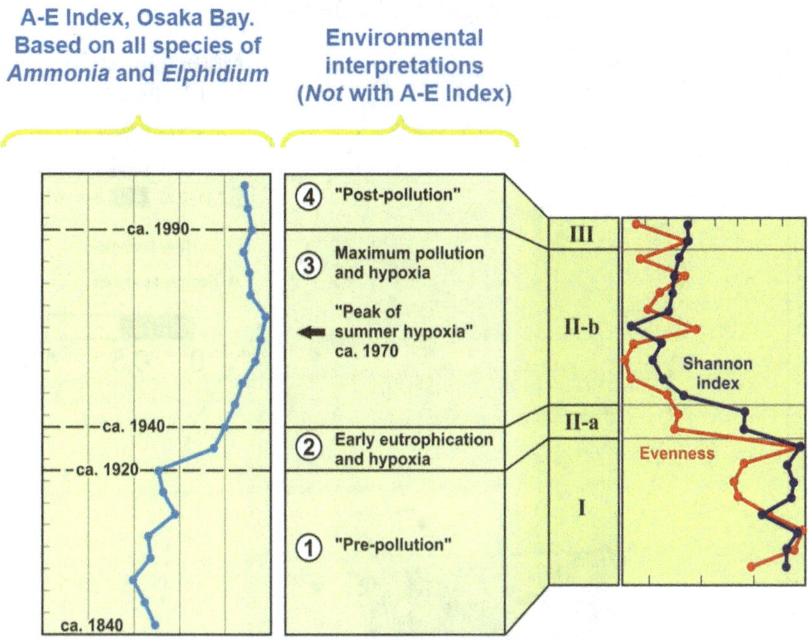


**Fig. 2.16** Eutrophication on shallow marine benthic over the last 150 years in Osaka Bay (Japan), found out four stages of pollution verified by Shannon index (diversity), evenness and AE index

Photosymbiont-bearing species (PSB) act as hosts to algal endosymbionts comparable to the zooxanthellae of corals, and the potential advantages of algal symbiosis to foraminifers lie in three major areas: energy from photosynthesis; enhancement of calcification; and uptake of host metabolites by symbiotic algae (Hallock 1999).

As with hermatypic corals, these large PSB prefer nutrient-poor, shallow, warm-water environments (Hallock 2000). The FI is based upon the hypothesis that species of benthic foraminifera that host algal endosymbionts prefer similar environmental conditions as hermatypic corals and can be used as proxies for reef and paleo reef health (Hallock et al. 2003).

The FORAM Index (FI) described by Hallock et al. (2003, 2012) is based on the proportions of three species groups: "symbiont-bearing," "opportunistic" or "stress-tolerant" (nomenclature changed in Carnahan et al. 2009), and "other small taxa."

If the proportions of these groups are Ps, Po, and Ph, respectively, then FI is given by $(10 \times Ps) + Po + (2 \times Ph)$, and an FI value of >4 indicates an environment "conducive to reef growth."

The index is weighted such that "to have an $FI > 2$, there must be some symbiont-bearing taxa, and for $FI > 4$, symbiont-bearing taxa must make up at least 25% of the assemblage" (Carnahan et al. 2009).

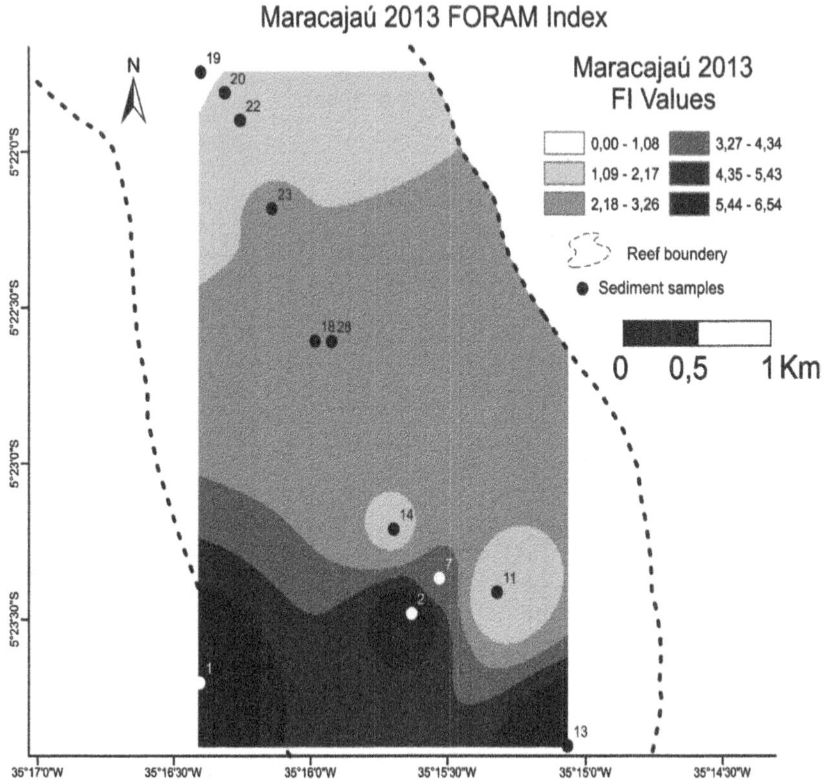


**Fig. 2.17** Gradient of F1 which higher values with environments conducive for reef growth are found in the south of the region. Lower values are on the north reef area is more vulnerable

A gradient of F1 values can be seen in Fig. 2.17, which higher F1 with environments conducive for reef growth occurs in the south of the region. Lower values are on the north of the reef and tend to decrease in their growth.

The PSB of Brazilian reefs used to calculate the FI index include seven species: *Amphistegina gibbosa, Amphisorus hemprichii, Archaias angulatus, Borelis schlumbergeri, Heterostegina antillarum), Peneroplis carinatus, and Laevipeneroplis proteus* and are displayed in Fig. 2.18.

- Paleo temperature index, a "Warm" species index

  Lutze (1977, 1979) recognized that certain deep-water benthic foraminiferal species were associated with 'warm' interglacial or 'cool' glacial conditions based on correlation with foraminiferal oxygen isotope records. Benthic foraminiferal species are then separated into a 'warm' group (*Laticarinina pauperata, Cibicidoides mundulus, C. robertsonianus, Cibicidoides* sp., *Hoeglundina elegans,* and *Bulimina aculeate)* versus a 'cold' group (*Pyrgo murrhina, Planulina wuellerstorfi, Uvigerina peregrina,* and *Globobulimina hoeglundi),* designated as 'warm' species index by Billups et al. (2020) from marine sediment samples from the Western Pacific Warm Pool (WPWP).

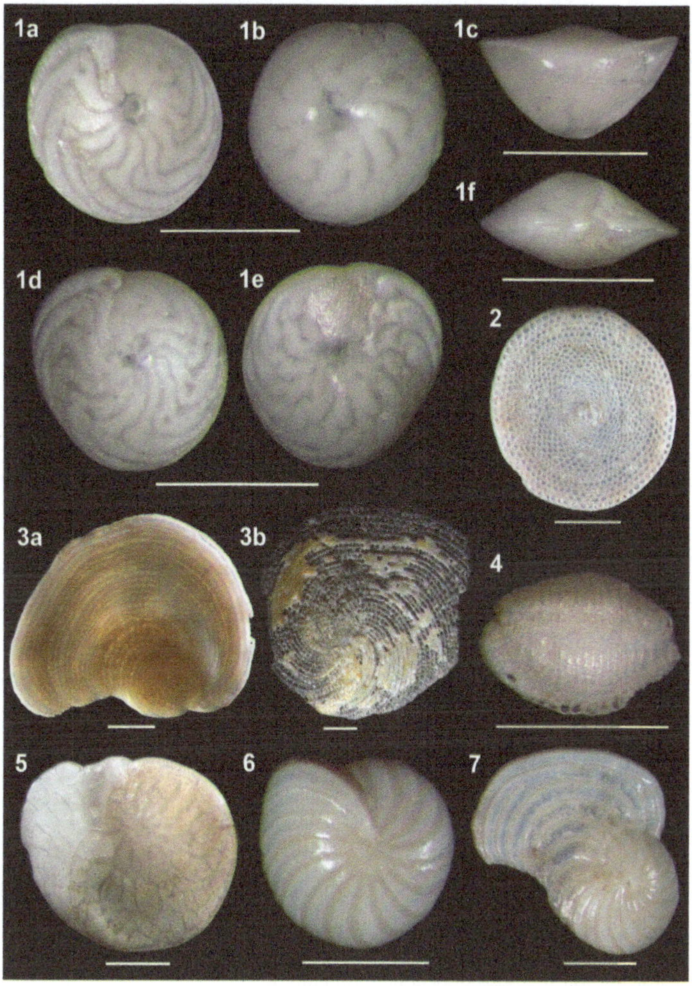


**Fig. 2.18** Original PSB of RN reefs used to calculate the FI index. *Amphistegina gibbosa* (1a–f), *Amphisorus hemprichii* (2), *Archaias angulatus* (3a, b. eroded), *Borelis schlumbergeri* (4), *Heterostegina antillarum* (5), *Peneroplis carinatus* (6), and *Laevipeneroplis proteus* (7)

This index is a robust tracer for temperature because it is based on a group of species rather than relative abundance changes of a single species.

There is association between 'warm' species index, organic carbon flux, and modern bottom water temperatures. Distribution of lower bathyal and upper abyssal can be separated into two distinct groups, one that reflects relatively warm (>3.5 °C) intermediate depth water with higher carbon flux (>3.5 g C $m^{-2}$ $year^{-1}$) versus a group that reflects the colder, deeper waters with lower carbon flux. These groups largely agree with the 'warm' versus 'cold' species

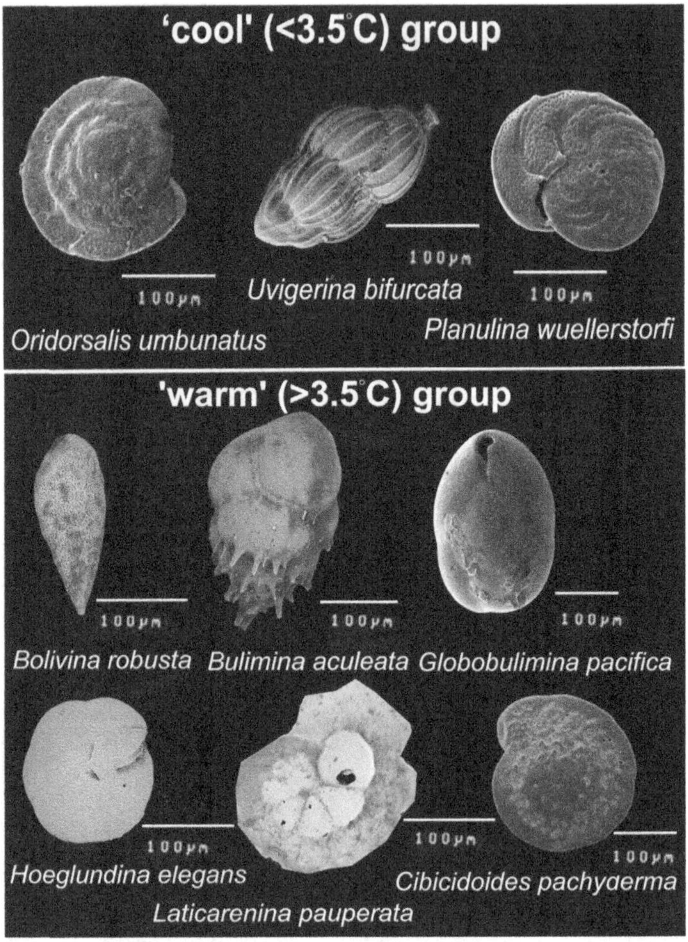


**Fig. 2.19** Cool and warm groups of benthic foraminifera

groups of Lutze (1979) and their approach in calculating a 'warm' species index (Fig. 2.19).

Large changes in the percent of 'warm' index is consistent with the patterns displayed by high-resolution benthic foraminiferal assemblages' changes over the past 0.5 Ma that relate to variations in carbon flux and productivity (Holbourn et al. 2005). At Papua New Guinea, down-core variations in the % 'warm' group remain relatively high consistent with high organic matter input via the Sepik River.

- Indicator species of paleo productivity

  Relatively high abundances of *Uvigerina proboscidea* reflect input of organic matter such as associated with down-slope movement in a tectonically unstable region. Fig. 2.20 shows variations between 'warm' species group and percentage

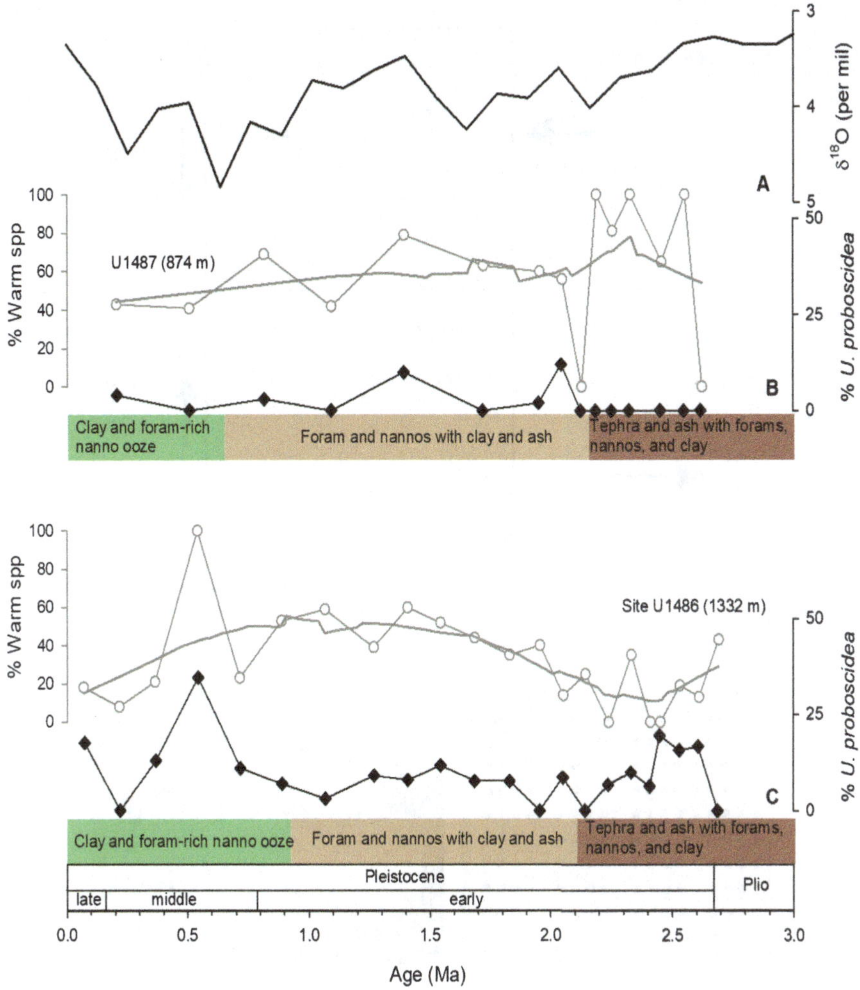


**Fig. 2.20** Variation between 'warm' species group and percent *U. proboscidea* as a tracer for carbon flux

of *U. proboscidea*, provides further evidence for the use of this index and the species itself as an independent tracer for carbon flux.

- Marine Stable Isotopic analyses and Paleoclimatology

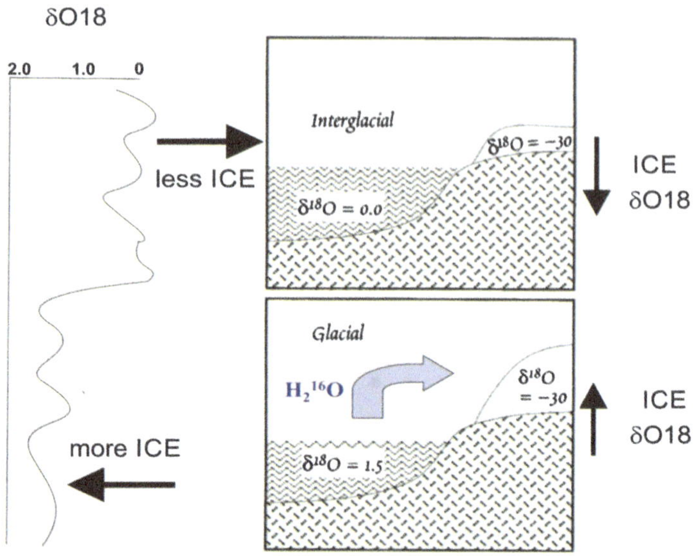


**Fig. 2.21** Relation of amount of ice and $\delta^{18}O$ of seawater in which calcification of the foraminifera shell took place

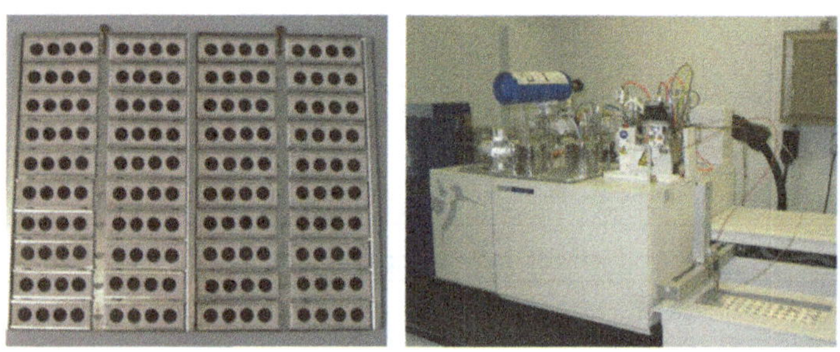

**Fig. 2.22** Samples prepared for stable isotopes analysis in the stable isotope mass spectrometer managed by K. Billups (University of Delaware)

## *Reading History Trapped in Shells*

The calcium carbonate ($CaCO_3$) in a foraminiferal shell contains O and C, which came from seawater at the time the creature was alive. They occur as a number of different isotopes which take part in reactions at slightly different rates. Therefore, processes as growth and decay of ice sheets during glacial-interglacial cycles, changes in the global and local temperature of seawater, and the growth and decomposition of organic matter, in the ocean and atmosphere affect the proportions of different isotopes (C and O) from Forams shells.

18Circulates in the main climatic subsystems, Fractionated whenever a phase transition between the reservoirs occurs. Controlled by mixing of runoff and seawater. Information on the temperature (amount of ice) from the isotopic composition ($\delta^{18}O$) of seawater in which calcification of the foraminifera shell took place (Fig. 2.21).

13Controlled by oxidation of organic matter, imported by rivers, Water mass circulation patterns and oceanic nutrient levels.

Reconstructions of paleotemperature, paleosalinity, sea-level fluctuations, paleoproductivity, and carbon export flux rates rely on assemblage composition and analyses of foraminiferal test geochemistry. Recent studies shown that conservative mixing controls the isotopic composition of water in bays and estuaries, and it is possible to use the marine stable isotopes ($\delta^{18}O$ and $\delta^{13}C$) values of benthic foraminifera to detect gradients in water-mass geochemistry.

Samples for this work were analysed in the stable isotope mass spectrometer of University of Delaware (Fig. 2.22).

To obtain a longer-term view is necessary to establish proxies that can track temporal changes in salinity or temperature. The marine stable isotopic composition of δ180 values from the *Elphidium poeyanum* shells was evaluated in the Bay in Delaware (USA) by Eichler et al. (2010) and their spatial distributions of stable isotope values are contoured for *E. poeyanum* in Fig. 2.23.

Low δ180 values for *Elphidium poeyanum* near freshwater sources (Love Creek and Herring Creek) and highest values near the marine inlet (Indian River Inlet). *E. poeyanum* reigns the colder, more saline outer bay near the marine inlet. Values of δ180 for unstained and stained tests of E. *poeyanum* correlate well with salinity

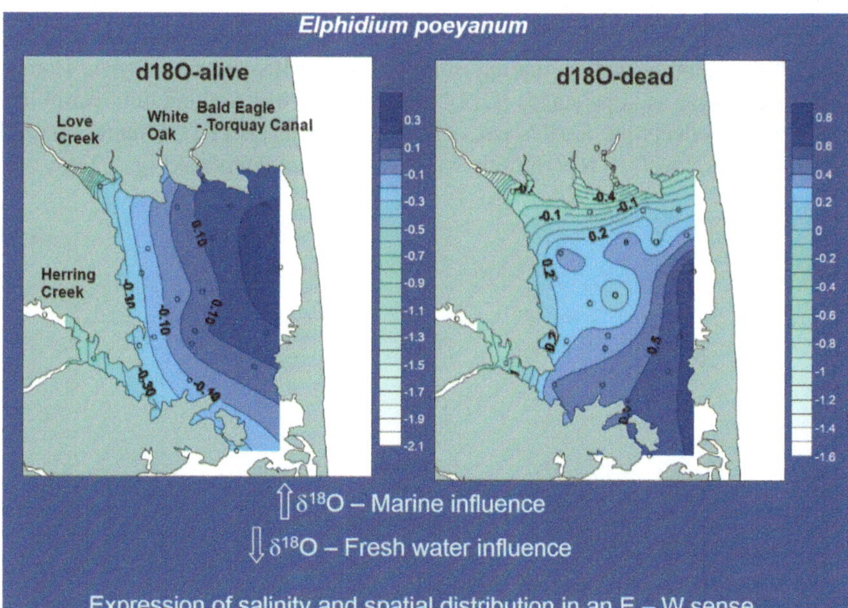


**Fig. 2.23** The stable isotopic composition of *Elphidium poeyanum*

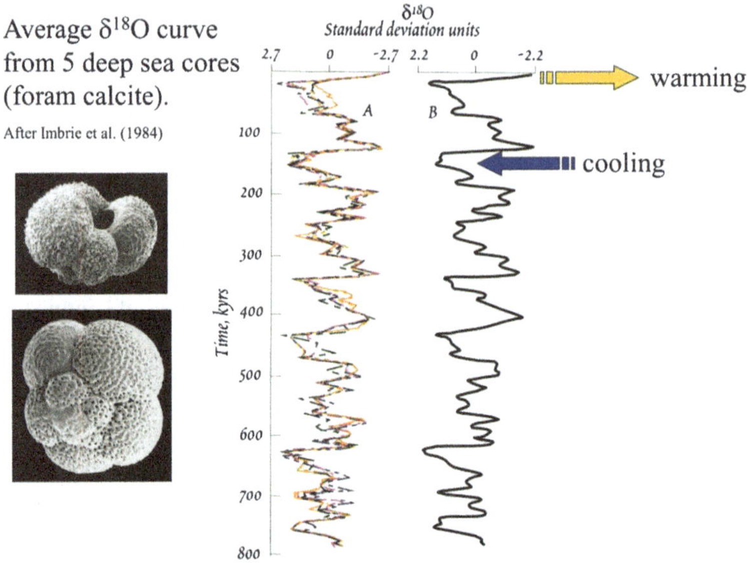


**Fig. 2.24** Example of a climate change during the last million year

in the recent marine sediments. Historical changes in the mixing between fresh versus marine waters of Rehoboth Bay in East and West spatial sense can be determined by analyzing faunal sequences and geochemical signatures in cores of its substratum.

From oxygen isotope data it is also possible to track climate change during the last million years, as well as marine isotope stages (MIS), marine oxygen-isotope stages, or oxygen isotope stages (OIS). These stages which alternating warm and cool periods reflects the Earth's paleoclimate. High levels of O18 represent cold glacial periods, low O18 representing warm interglacial intervals (Fig. 2.24).

# Chapter 3
# Microfossil Shells Are Carbon Story Tellers: Microfossil Communities: First Responders to Environmental Impacts

**Abstract** Most microfossil organisms are visible only under a microscope and only a few of them are the cause of diseases. We all depend on microorganisms in the environment, because they recycle matter in nature, transforming complex compounds into simpler ones readily absorbed by others. "The storytellers of space and time" have particularities and habits, which tell pieces of the environmental puzzle in the present time and that of the past, because they respond quickly to changes in abiotic variables (temperature, oxygenation, pH, geoproperties) linked to abrupt cooling or warming in oceans worldwide. Extinction and recovery events with little migratory power recorded by microfossils illustrates the role of climate and oceanographic changes that drive the short and long-term successional changes in ocean ecosystems. Their sensitivity to natural changes, pollution or contamination, and the occurrence or absence of opportunistic species unravels dynamic and interesting data. This data shows that microfossil communities inter and intra competition patterns exist temporally and spatially, and is easily recorded in any environmental disturbancc. These tiny beautifully articulated shells, of the thecamoebian "Freshwater Finders", ostracode, the crustacean group of "predators in space and time", nanofossils "the tiniest paleo watches", radiolarians "the delicate sedimentary rock fairytale", and the diatoms as "The beautifully done silicate factory", and the last but not the least, the planktonic and benthic foraminifera as the "Glacial and interglacial calcareous factories" are all different actors on the stage. Planktonic are the "Big Ben of time", and benthic are the "Small giants of show and tell in space and time". All microfossils shells are proxies that provide opportunities to evaluate cycles of the "changing world" of marine communities today and into past millions of years. They clearly explain climate and environment since their abundance in the sediment corresponds to past fluctuations over time.

**Keywords** Microfossils · Bio indicators · Climate change · Oxygenation · Vulnerability · Mitigation · Adaptation



P. P. Beck Eichler, C. P. Barker, *Benthic Foraminiferal Ecology*,
https://doi.org/10.1007/978-3-030-61463-8_3

## Introduction

Studies worldwide using comprehensive quantitative analysis of different microfossils with high-resolution records respond to glacial and interglacial episodes. However, what are Microfossils? They are fossils visible only under a microscope. They can be whole shells of microscopic animals such as benthic and planktonic foraminifera, thecamoebian, ostracode, nanofossil, radiolarians, diatoms, and micro mollusks. They can also be small parts of macroscopic animals, spicules of sponges, fish teeth, or reproductive parts of plants and fungi such as pollens and spores or even isolated cells, or in-group forming filaments and networks of bacteria.

Of the thousands of known microorganisms, only a small number can cause diseases. All animals and plants depend on chemical transformations carried out by microorganisms in the environment, and microbes promote the recycling of matter in nature, transforming complex compounds into simpler ones readily absorbed by others. This chapter will demonstrate what seafloor the microfossil community of thecamoebian, ostracode, nanofossil, radiolarians, diatoms, planktonic and benthic foraminifera have that makes them "the storytellers of space and time". This includes their particularities and habits, which can tell a piece of the environmental puzzle in the present time and that of the past. They respond to changes in the abiotic variables, temperature, oxygenation, pH, and geo properties, and are associated with abrupt cooling or warming in oceans worldwide linked to climate changes. This is because the extinction and recovery events seen within the microfossil-record illustrates the role of climate and oceanographic changes that drive the short and long-term successional changes in ocean ecosystems.

Micropaleontology, the study of microfossils, helps us to characterize recent and older environments by selecting environmental indicator microorganisms, also known as bio indicators, by using high and low resolution data. As bio indicators these microfossils have to; be ubiquitous in the marine environment, present wide taxonomic diversity with short reproductive cycles, and have resistant shells that can be preserved. They should be small, abundant and easily collected with little migratory power. They have to be specific to environmental conditions, be sensitive to either natural, pollution or contamination changes. They respond to these changes in many ways; through deformations in the shells, by the occurrence of opportunistic species (dominance of some indicator species), or by the absence of species (decreased diversity).

Here we present the important, dynamic and interesting communities of microfossils where naturally inter and intra competition patterns exist temporally and spatially, and where it is traceable if there is any environmental disturbance. For example, episodes of increase in water temperature, decrease in oxygenation, and abrupt climatic warming provide opportunities to assess the effects of changing oxygenation on benthic communities.

Anthropogenic climate change is likely to decrease oceanic oxygen ($O_2$) concentrations, with wide range and often devastating effects on whole communities of marine ecosystems. Our knowledge even if it is largely restricted to local investigations worldwide, is linked to abrupt ecosystem-scale responses and climate-forced deoxygenation, and can pinpoint recent areas that are more environmentally vulnerable based on past communities which then allows us to recommend mitigation and adaptation strategies for communities, municipalities and industry.

## Presenting Microfossils as "Story Tellers"

The tiny shells of thecamoebian, ostracode, nanofossil, radiolarians, diatoms, planktonic, and benthic foraminifera provide opportunities to assess the effects of a "changing world" on marine communities millions of years into the past. Figure 3.1

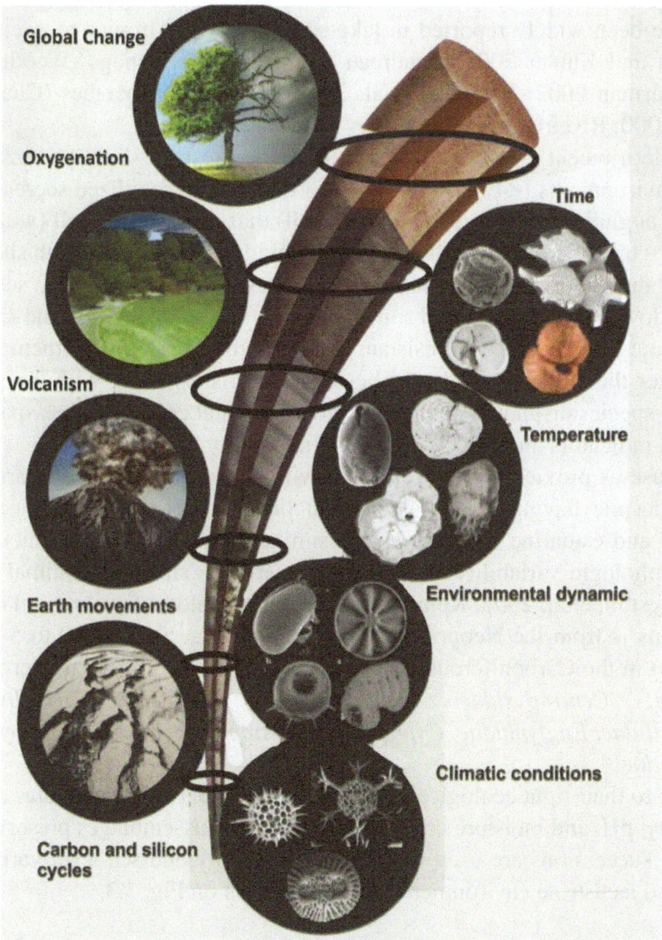


**Fig. 3.1** Microfossils and environmental applicability

illustrates this assessment and we hope that you will enjoy learning a bit about each microfossil, its description and environmental applicability on global change. These include oxygenation, volcanism, earth movements, and carbon and silicon cycles in time, and what temperature is dependent on the environmental dynamics and how can we tell climatic conditions?

## *Thecamoebian as "Freshwater Finders"*

Thecamoebian (Arcellaceans) are unicellular eukaryotic heterotrophic Protista (Phylum Amoebozoa) shelled amoebae (testate forms) that occur widely in a variety of freshwater habitats (Loeblich et al. 1964; Medioli and Scott 1983, 1988), and in marginally brackish environments (Charman 2002; Patterson and Kumar 2002). They have been widely reported in lake sediments from tropics to Arctic regions (Patterson and Kumar 2002; Boudreau et al. 2005), peat bogs (Woodland et al. 1998; Charman 2002; Swindles et al. 2010), and in salt marshes (Charman and Hendon 2000; Riveiros et al. 2007).

In modern recent environments, they colonize fresh to slightly brackish water aquatic environments (<4 PSU) and they form a simple fossilized secreted (autogenous) or agglutinated (exogenous) test (shell) that encloses the cell (with an aperture from which the pseudopodia emerge), providing the amoeba with shelter from predators and environmental conditions. This test is preserved in the sedimentary record following their death, and some examples of *Centropyxidae* and *Difflugidae* are illustrate on Fig. 3.2. The resistant nature of their tests allows them to be preserved after the amoeba has died. These characteristics along with the sensitivity that some species display as changes in environmental conditions has sparked their use as bio indicators and proxies of paleoclimatology in recent years.

Their use as proxies for a variety of environmental and climatic parameters is because despite having a long geological history, thecamoebian lineages from lacustrine and estuarine successions are similar to modern forms which suggests little morphologic variability in the group over time showing minimal evolution (Van Hengstum et al. 2007; Kumar et al. 2011). The oldest fossil record of the thecamoebians is from the Neoproterozoic (geologic time from 1000 to 541 million years ago) in the Carboniferous, and they are represented by 11 modern families: *Arcellidae, Centropyxidae, Plagiopyxidae, Difflugidae, Hyalophenidae, Phyrganellidae, Euglyphidae, Cyphoderiidae, Amphitremidae, Trigonopyxidae* and *Trinematidae*.

Owing to their tight ecological zonation with respect to temperature, conductivity, salinity, pH, and moisture content, thecamoebian assemblages preserved in late Holocene successions are used to reconstruct salinity and sea-level variations, on coastal and lacustrine environments, see illustration on Fig. 3.3.

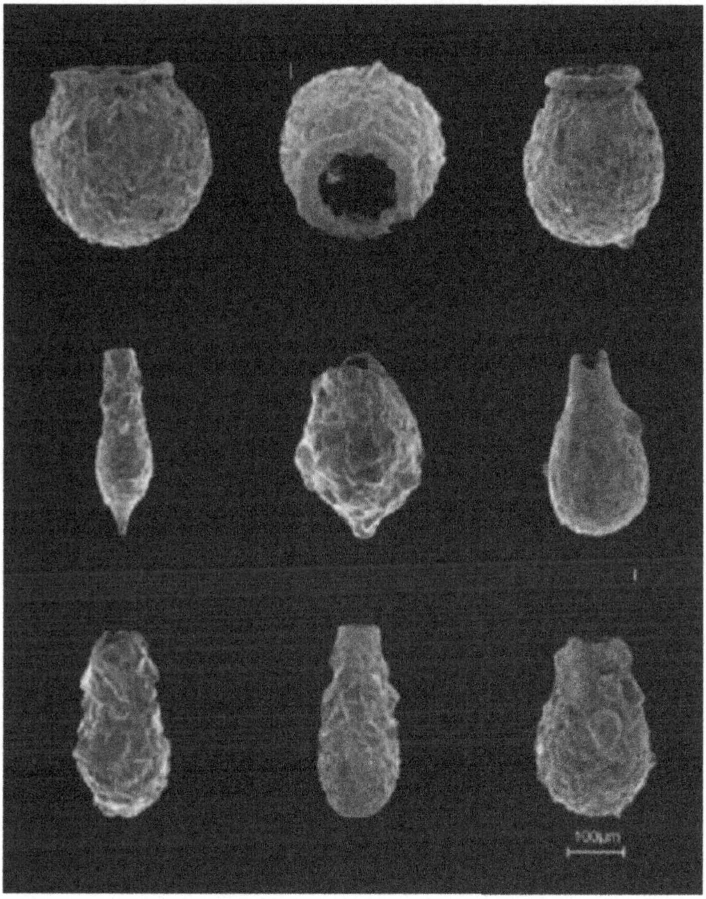


**Fig. 3.2** Centropyxidae and Difflugidae

## *Ostracods as "Predators in Space and Time"*

Ecologically, marine ostracods (Phylum Arthropoda, class Ostracoda) are part of the zooplankton or benthos, living in or on the top layer of the seabed. Some can be found in freshwater (Podocopids), and moist forest soils (Mesocypris). They have a wide range of diets, and the group includes carnivores, herbivores, scavengers and filter feeders. These little monsters eat it all!

They are one of the most successful crustaceans' groups known as "mussel shrimp" or "seed shrimp". Only 13,000 are current living species and the paleo archive counts 65,000 identified species. Their small bodies of 1 mm in length are flattened from side to side protected by a bivalve calcareous shell that encloses the entire body. The two valves are located in the upper (dorsal) region of the body,

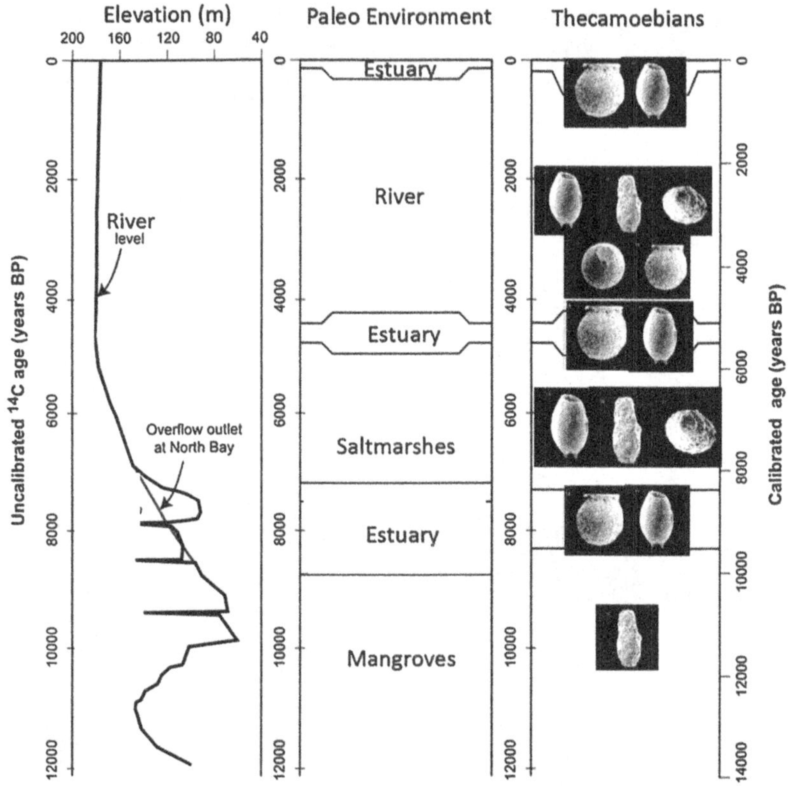


Fig. 3.3 Shows an environmental reconstruction of 14,000 years BP using Thecamoebians communities and paleo

typical five pairs of appendages are located on their heads but only 1–3 pairs of appendages on the rest of the body, and they shed the carapace with each molt.

Most ostracods are found crawling on or burrowing into the sediments at the bottom of the ocean or lakes. A zonation based on the ornamentation of their bodies and the salinity is illustrate on Fig. 3.4. In a general way, the more ornamented the test more saline is the environment.

Some species are predators; others crawl in moist terrestrial habitats such as mosses where they feed, on dead organic material, suspended organic particles, or microscopic plants. The ostracods have the most complete fossil record of any of the crustaceans found from Cambrian to the present. Useful for the bio zoning of marine strata on a local or regional scale, ostracods are valuable indicators of paleo environments because of their widespread occurrence in coastal zones, small size, and are easily preserved. Calcified bivalve carapaces are commonly found in records of Microfossils and easily identified and linked to any particular environment. Figure 3.5 illustrates species of ostracods.

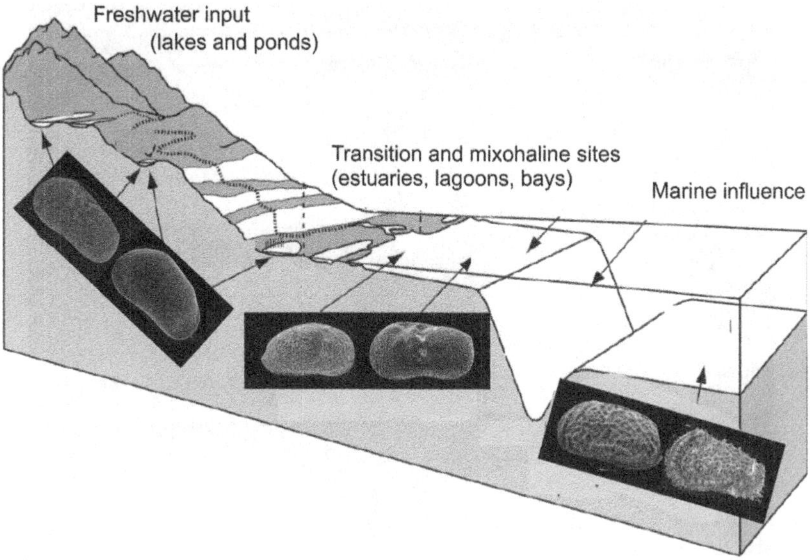


**Fig. 3.4** Ostracods zonation related mainly to salinity and bodies' ornamentation found in the sediments at the bottom of the ocean or lakes. From less saline, and less ornamented test to more saline and more tests' ornamentation

## *Calcareous Nano fossil "the Tiniest Paleo Watches"*

Fossil coccoliths (disc-like form) and nannoliths (variously shaped form) of algal origin constitute the smallest group of calcareous Nano fossils, and they usually measure less than 30 μm in size (Fig. 3.6). The most common members of the group are the remains of haptophyte algae or, more specifically, the coccolith-bearing subgroup of haptophytes, the coccolithophores.

Chronologically as described in the fossil record, calcareous Nano fossils first appeared in the Late Triassic, where one of the five major mass extinction events of the Earth event began. The corresponding series of their appearance on the fossil record spans the time between 237 and 201.3 Ma (million years ago) and was restricted to low latitudes.

The presence and absence of the Nano group are very accurate to tell time. They are the "watches of the past" in the Triassic/Jurassic boundary extinction event, which have claimed all, but only one living Nano fossil species (*Crucirhabdulus primulas*) for example. Throughout the Jurassic and Cretaceous periods, Nano fossils reestablished their dominance within the oceans and colonized all marine environments, and by the Late Cretaceous when they reached a diversity peak. However, they were affected by the Cretaceous/Tertiary (K/T) boundary extinction event during which over 90% of the species became extinct. This influx was followed by a significant decline that continued into and within the Oligocene. The Miocene was

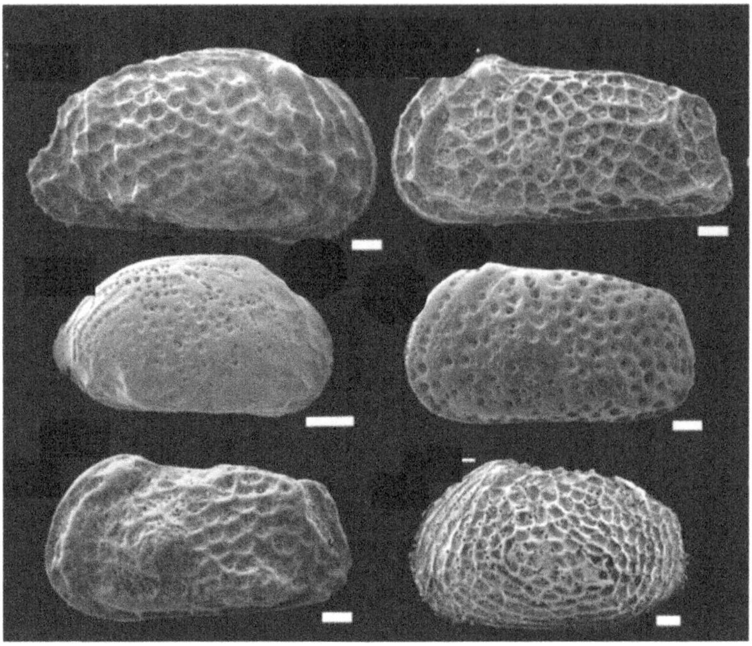

**Fig. 3.5** Different species of Ostracoda and its different ornamentations, which identify them as time and space

**Fig. 3,6** Calcareous Nano fossils species

a period of significant increase in Nano fossil diversity, which again gave way to a more pronounced increase that continued into the Recent.

As pointed out, calcareous Nano fossils are excellent proxies useful in the biostratigraphy of marine sediments of Jurassic to Recent ages. This is because they are abundant, rapidly evolving, and cosmopolitan. In addition, their abundance and very small sizes means that they are statistically studied. Furthermore, the preparation technique is quick, uncomplicated, and does not involve the use of dangerous chemicals. The simple preparation technique has particularly earned calcareous Nano fossil study a preferred status in well-site biostratigraphy.

## *Radiolarians "the Delicate Sedimentary Rock Fairytale"*

Radiolarians are a variety of tiny exquisitely beautifully sculptured structured skeletons of remarkable delicacy, and great spherically symmetrical complexity characteristic of most modern species. Their beauty is shown in Fig. 3.7.

This protozoan of the class Polycystinea (superclass Actinopoda) occur abundantly in the upper layers of major oceanic sites worldwide, but only the silica elaborately perforated skeleton is preserved under specific conditions across geographic regions. Exclusively open ocean lives on the plankton, with some species limited to certain regions and serve as indicators of water mass properties such as temperature, salinity, and total biological productivity. The test has a variety of patterns, forming a series either like plates or of loose needle-shaped spicules. Pseudopodia extend through the perforated skeleton. The skeletal remains of radiolarians settle to the ocean floor and form radiolarian ooze. When the ocean bottom is lifted and transformed into land, the ooze becomes sedimentary rock. Silica deposits, such as flint, and the abrasive Tripoli, originated from radiolarian skeletons. Fossil radiolarians have been found that date to Pre Cambrian Time (3.96 billion to 540 million years ago).

These microfossils are an important source of data in biostratigraphic and paleoclimate studies. The Radiolarian fossil record shows periods where Phanerozoic oceans existed and they are employed with great success in biological evolution and climate change. Radiolarian biostratigraphy dating of Cenozoic deep-sea sediments show speciation and factors controlling macro evolutionary change in entire faunas, and they have been employed to understand productivity change in the oceans, using the ratio of surface to deep-water species as an organic carbon export indicator. Radiolarian research, while successful, is still limited by poor knowledge of the biology of the group, and by incomplete taxonomy, even at the most basic level of species descriptions.

Statistical analysis of micro paleontological data on radiolarians in the bottom surface sediments are consistent and the biogeographic zonation are in agreement with oceanographic setting. There are clear temperature limits between assemblages. Distribution and interaction of water masses of different origin—from the cold Polar to warm North Atlantic is reflected in the occurrence of radiolarian assemblages in all oceans. Variations in the number and kind of radiolarian species (based on skeletal form) in relation to depth in the sediment provide information about climatic and environmental conditions in the overlying water mass at the time

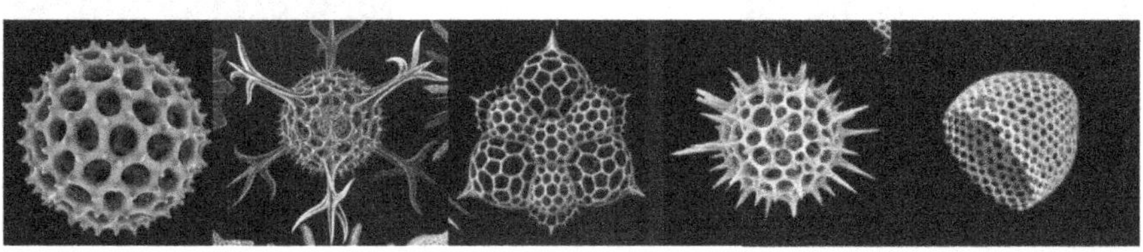

**Fig. 3.7** Calcareous Radiolarians species

the radiolarian skeletons were deposited at that geographic location. The radiolarians are second only to diatomacea as a major source of biogenic opal (silicate) deposited in the ocean sediments.

## *Diatomacea "The Beautifully Done Silicate Factory"*

Diatom (class Bacillariophyceae) with about 16,000 species are the most species-rich group of algae found in sediments or attached to solid substances in all the waters of Earth and are very beautiful in its forms (Fig. 3.8).

Diatoms are among the most important microscopic sea organisms and serve directly or indirectly as food for many animals. Paleo limnological diatom data comprise counts of many species expressed as percentages for each sample, and pH reconstructions of past lake-water pH, and rate of change are possible by using this microfossil. They are ecologically widespread and have global significance in the carbon and silicon cycles, and are used in ecological monitoring, paleo ecological reconstruction, and stratigraphic correlation.

Several model systems have provided valuable insights into the nature of diatom species, because many species contain several reproductively isolated entities that are worth taxonomic recognition at species level. Phenotypic plasticity, although present, is not a serious problem for diatom taxonomy.

The examination of diatom community structure is centered on local environmental factors instead of broad-scale biogeographical factors largely because diatom distribution has been considered cosmopolitan. Endemism has been seriously underestimated among diatoms, and biogeographical and stratigraphic patterns are difficult to discern because of the lack of trustworthy data and because the taxo-

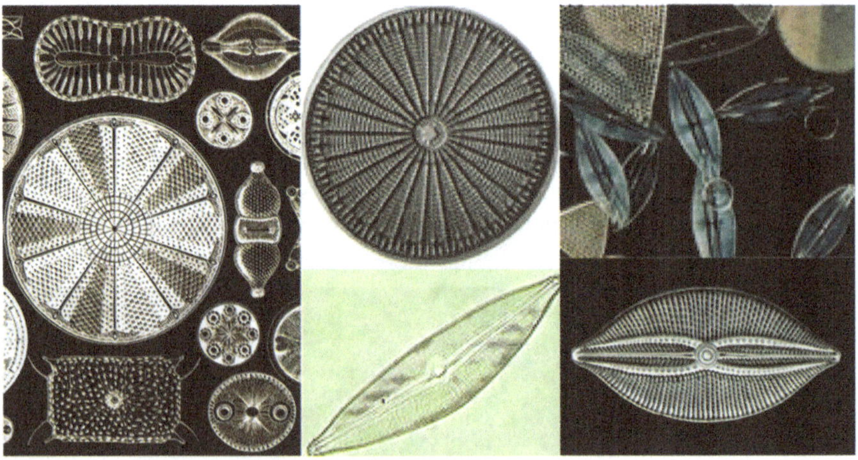

**Fig. 3.8** Examples of diatoms

nomic concepts of many authors are undocumented. However, a number of recent publications have stressed the role of regional constraints such as historical factors, evolution and variability in modern day dispersal ability in determining the diatom community structure.

Freshwater diatom communities are strongly spatially structured. According to studies from lakes and streams, spatial factors account for about 20–30% of the total explainable community variation, suggesting that diatoms lack strict ubiquitous dispersal. Assessment programs utilizing diatoms should consider spatial factors using for example, regional stratification as diatom communities are strongly spatially structured.

## *Planktonic and Benthic Foraminifera the "Glacial and Interglacial Calcareous Factories"*

Marine sediment cores covering the past climatic cycles are investigated to elucidate climate changes with planktonic and benthic foraminiferal shells as proxies, since foraminiferal specimens per gram of sediment corresponds to past fluctuations over time. Highest abundances occur during peaks in interglacial periods, while reduced number of foraminiferal tests marks glacial periods. In a general view, the relative proportion of planktonic and benthic foraminiferal abundance seems to have varied spatially similarly between each inter glaciation and short time intervals.

Properties of Foraminiferal shells that can be used for Paleo proxies are census counts of taxa, ecological or functional types, specific species presence or absence, semi-quantitative abundance estimates, absolute and relative abundances, biometry, measurements made on individual specimens, morphology, size, shape shell ultrastructure, color and weight, taphonomic process (alterations to fossils incurred after deposition), preservation and ecological indices, and fragmentation.

High-resolution studies of the past unravels that benthic and pelagic planktonic foraminifer increased in number after the end of last glaciation, linking changes in postglacial water masses to a direct impact on sea-surface and bottom productivity. This direct coupling between pelagic and benthic faunal productivity persists, although with some notable variations, throughout the Holocene (Anthropocene).

High correlation coefficient between surface temperature water and the most dominant benthic suspension feeder of a determined site is a strong pelagic-benthic coupling evidence of a continuous vertical connection of surface and bottom habitats through time. In a general way, variations in temperature of upper waters, is responsible for changes in the planktonic foraminiferal assemblage, while changes in nutrient availability and substrate oxygenation, because of input variations from different source areas at the precession-scale cycles (about 26,000 years) affect the benthic foraminiferal assemblage.

## *Planktonic Foraminifera "the Big Ben of Time"*

The use of planktonic foraminifera as tracers of ocean properties is a well-developed field of science. Production and excellent preservation of planktonic foraminifera has enable Paleoceanography develop so well in foraminiferal fossils from oceanic sediments, producing probably the best fossil record on Earth, with enormous archives of morphological change, faunal variations, and habitat characteristics (Fig. 3.9).

Their properties of fossil assemblage's rates and patterns of evolution or as a substrate for extraction of geochemical signals are wide and helps answer biostratigraphy and climate questions worldwide. Planktonic foraminifera plays a central role in paleoceanography providing science with robust and reliable proxies. These inconspicuous organisms and their tiny shells are the real heroes of our endeavor to unravel the past of our planet.

Foraminiferal shells are responsible for deposition of a large portion of deep-sea biogenic carbonate. Over a period of 500 years planktonic foraminifera deposit a mass of carbon equal to the entire biosphere!!

Murray (1973), who recognized that extant species in the plankton and in sea floor sediments are distributed in global belts related to surface-water temperatures, was the one to first describe the potential for planktonic foraminifera as tracers of surface-water properties. Later on, Schott (1935) pioneered the use of quantitative census counts and discovered that fossil assemblages in short deep-sea cores changed between glacial and interglacial times.

The distribution and abundance of planktonic foraminifera species is strongly linked to surface-water properties since they inhabit the water column. Sea-surface temperature (SST) appears to be the single most important factor controlling assem-

**Fig. 3.9** Planktonic species

blage composition (Morey et al. 2005), diversity (Rutherford et al. 1999), and shell size (Schmidt et al. 2004b). Laboratory experiments (Bijma et al. 1990b) and sediment-trap observations (Zaric et al. 2005) indicate that planktonic foraminifera species survive under a large range of SST, but that their optimum ranges are typically narrow and distinct.

Reconstructions of Pleistocene climate variation uses planktonic foraminifera since the beginning of paleoceanography, and the value of foraminiferal calcite as a recorder of chemical and isotopic signals was recognized by Emiliani (1954a, b, 1956). Stable isotopic signals extracted from planktonic foraminifera are an accurate tool for recognition of glacial cycles, and of orbital pacing of the ice ages (Shackleton and Opdyke 1973; Hays et al. 1976). The chemical composition of foraminiferal calcite proved to be useful for the development of proxies, especially to reconstruct past seawater chemistry and biogeochemical cycles using stable-isotopic and geochemical proxies, for example.

Planktonic foraminifera are marine heterotrophic, omnivore to selective carnivore, that surround their unicellular cytoplasmatic body with elaborate calcite shells containing eukaryotic cellular organelles and fibrillary bodies, unique features to planktonic foraminifera that act to control buoyancy (Hemleben et al. 2012). The external network serves to collect food particles and transport it toward the primary opening of the shell (aperture, foramen). Herbivorous and omnivorous species consume phytoplankton, mainly diatoms and dinoflagellates, while carnivorous species prey on copepods, ciliates, and other similarly sized zooplankton (Hemleben et al. 2012).

Sexual reproduction has to be synchronized in space and time in order to maximize the chances of gametes from different individuals finding each other…

Therefore most shallow-water species appear to reproduce in pace with the synodic lunar cycle (*Hastigerina pelagica, Globigerinoides sacculifer, Globigerina bulloides*) or half-synodic lunar cycle (*Globigerinoides ruber*) (Spindler et al. 1979; Bijma et al. 1990a; Schiebel et al. 1997), and lunar pacing appears important for carbonate production in the tropical oceans (Kawahata et al. 2002). Deep-dwelling species like *Globorotalia truncatulinoides* may follow longer, perhaps yearly, reproductive cycles.

Species of planktonic foraminifera are grouped into five main assemblages that define the tropical, subtropical, temperate, subpolar, and polar provinces (Bradshaw 1959; AWH & Tolderlund 1971). The warm-water provinces cover almost two-thirds of the world oceans. Most are cosmopolitan within their preferred bio province, although three Indopacific (*Globigerinella adamsi, Globoquadrina conglomerata, Globorotaloides hexagonus*) and one Atlantic tropical species (*Globigerina ruber*) are modern and endemic (Fig. 3.10).

In modern oceans, the polar waters of both hemispheres are dominated by a single small species, while the highest diversity and largest sizes are found in the oligotrophic subtropical gyres. The increase in SST toward the equator accompanies a proportional increase in surface-water stratification. The strength of vertical gradients in the ocean determines the number of physical niches available for passively floating plankton, and may thus control their diversity and morphological

**Fig. 3.10** *Globigerina ruber* (in pink) from deep ocean marine sediments

disparity (as reflected by size range) (Schmidt et al. 2004a, b). The general trend toward higher diversity and larger sizes with increasing SST is reversed in equatorial and coastal upwelling zone with higher population densities of smaller species.

Large, symbiont-bearing carnivorous specialists are adapted to oligotrophic conditions, and in high-productivity regimes, they are outnumbered by omnivorous and herbivorous species such as *Globigerina bulloides* and *Globigerinita glutinata*. The distribution of the provinces in the modern ocean follows sea-surface temperature gradients, reflecting strong relationship between SST and species abundances. These opportunists can rapidly react to organic particle redistribution and phytoplankton blooms after nutrient entrapment (Schiebel et al. 1995).

Oscillatory pulses of primary productivity, coupled with seasonal sea-surface temperature cycle result in predictable successions of planktonic foraminifera species, which react to changes in the environment according to their ecological preferences. The flux rate of planktonic foraminiferal shells follows primary productivity cycles with a lag of several week where polar oceans, the flux peak is observed during summer, and in temperate oceans the spring flux maximum is often followed by a smaller autumn peak.

Tropical and subtropical oceans are characterized by a steady rain of foraminiferal shells throughout the year. Within the range of normal marine conditions (33–36%), salinity does not appear to significantly influence on planktonic foraminifera (Hemleben et al. 2012) however in modern oceans, no planktonic foraminifera lives under hypo saline conditions.

Many properties of individual organisms and ecological systems are affected by physical and chemical parameters of their habitat. If we knew how the environment modifies the basic genetic design of organisms and how it controls their spatial and temporal distribution, we could use the fossil record of such organisms to reconstruct state and variation of past environments. The various signals locked in fossils are rarely related to individual environmental parameters but rather to an entire community of species. Therefore, paleo environmental reconstructions rely on recipes and algorithms describing ways of how to relate measurements and observations made on fossils and other geological material to past environmental variables. Because of that, physical and chemical properties of foraminiferal shells provide a multitude of paleo proxies, based on the chemical composition and morphology of their shells as well as species abundance patterns.

Understanding most paleo proxies is rare, particularly of those based on fossils and their physical form. As soon as life with all its complexities enters the equation, paleo oceanography had to resort to the process of empirical calibration. The relationship between a physical parameter of a fossil and an environmental variable is derived by observing and describing the distribution of taxa and their properties in the present-day ocean. Therefore, physical and chemical properties of foraminiferal shells are easily determined with high precision following different taphonomic pathways and proxies from the same fossil assemblage, providing a unique opportunity to assess the robustness of such paleo environmental reconstructions.

Methodologically this process is relatively simple but involves a number of assumptions that limit the applicability of each empirically calibrated proxy. An empirical calibration requires a database of measurements or recordings of a physical parameter of an extant organism, with simultaneous observations of the desired environmental variables associated with its habitat and statistical analyses to determine the form of the relationship between the two types of data. The main assumptions of this process are that the target environmental variable exerts a significant influence on the measured property, and that the mathematical tool is able to describe this relationship effectively.

Additional questions that need to be considered when applying an empirical calibration to a fossil sample: how far back in time can we go? How can we accurately recognize that a proxy is yielding reconstructions of reliability? Both issues reflect the stationarity principle of empirical calibration. This principle states that fossil's properties, relationships among them, and the environment must remain the same throughout the time duration of the proxy application. The exact range in time of each calibration depends on the rate of evolution in the group of fossils knowing that reliability is highest in samples derived from the same period.

An ecological calibration based on qualitative changes of proxy signs on modern planktonic foraminifera are used with high confidence, during the last glacial cycle, because taxonomic uniformity of Cenozoic planktonic foraminifera has led to calibrated proxies in older times (Andersson 1997) delivering robust qualitative information (warmer, colder) that can be applied further back in time. Figure 3.11 shows an example of determined species used as bio indicators of Periods, Epoch, and Ages in time.

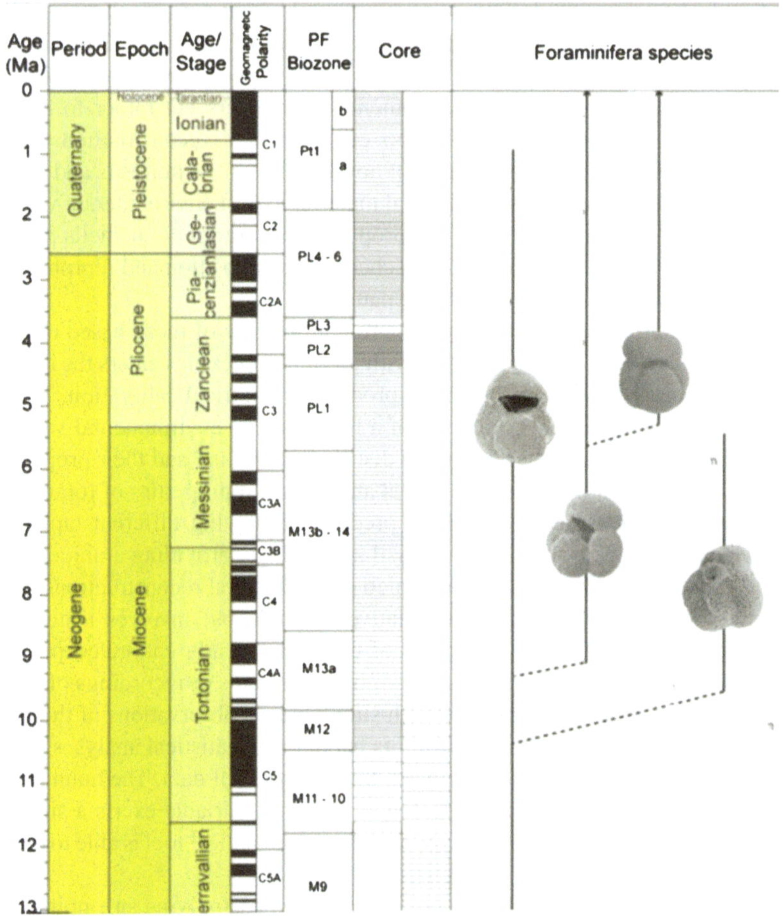


**Fig. 3.11** An example of an output based on planktonic foraminiferal species, telling age and bio zones in marine sediment cores in the last millions of years

## Indicator Species and Assemblages

The simplest types of foraminiferal proxies are based on abundances of ecologically significant (indicator) species. For example, the deep-dwelling *Globorotalia truncatulinoides* exhibits a yearly reproductive cycle during which it descends to considerable depth in the ocean (Schiebel and Hemleben 2005). It can only complete its reproductive cycle if the waters it passes through are oxygenated.

The *Globorotalia scitula* in sub-thermocline waters rich in organic debris allowed Schiebel, Schmuker, Alves, and Hemleben (2002) to trace the position of the Azores Front during the late Pleistocene. The most commonly used index species is the shallow-dwelling opportunistic *Globigerina bulloides*, which thrives in high-productivity environments, making it a good indicator of upwelling intensity

(Thiede 1975; Conan et al. 2002). Gupta et al. (2003) used the abundance of this species to reconstruct Milankovitch oscillations (collective effects of changes in the Earth's movements on its climate over thousands of years) in Holocene monsoon-driven upwelling in the Arabian Sea.

Grouping ecologically similar species into "indicator assemblages" provides a more robust approach. The proportion between warm and cold-water species could be used to track the thermal history of the ocean's surface on geological time scales (Ingle 1977), as well as with centennial resolution (Rohling, Mayewski, Abu-Zied, Casford, Hayes 2002).

A specific fauna is known to be associated with the warm Agulhas Current off South Africa (Giraudeau 1993) and past changes in the abundance of this "Agulhas fauna" can be used to reconstruct the history of surface-water exchange between the Indian and Atlantic Oceans on glacial/interglacial time scales (Peeters et al. 2004). In the North Atlantic, planktonic foraminiferal assemblages allow an indirect reconstruction of sea-ice distribution. Kucera et al. (2005) showed that surface sediment samples deposited below the modern Arctic Domain surface water in the North Atlantic always contain a small but detectable fraction (0.3–1%) of sub-polar species whereas in the perennially ice-covered regions these sub-polar species are totally absent. Presence of small sub-polar species are proxy of seasonally ice-free conditions.

### Coiling Direction

Planktonic foraminiferal shells with trochospirally arranged chambers can exhibit either dextral (right-handed) or sinistral (left-handed) coiling. Some species show a strong preference for right-handed, or left-handed coiling, while other species exhibit mixed coiling proportions. The ratio between the two coiling types of a species can vary through time and or space (Fig. 3.12).

Patterns of distinct shifts in coiling preference have been recognized and quantitatively characterized among several species of modern planktonic foraminifera. Assuming that coiling direction in such species reflected ecophenotypic response to environmental parameters, mainly sea-surface temperature (Ericson 1959; Ewing et al. 1954; Boltovskoy 1973), coiling ratios became an important early tool to reconstruct past marine environments. The determination of coiling ratios is rapid, accurate, and reproducible. A census of 50–100 specimens is normally sufficient to detect environmentally significant changes in coiling direction. A coiling direction proxy is based on the remarkably strong and consistent relationship between coiling direction and sea-surface temperature in high-latitude species.

It is important to remind you about the main factors affecting the incorporation of geochemical signatures into foraminiferal shells. The chemical composition of foraminiferal shells derives from the chemistry of the ambient seawater. Proxies that are known to be incorporated into the shells require information about habitat and ecology of the analyzed species. The fossil assemblage is biased toward shells deposited during the time of maximum production. Depending on the species, and

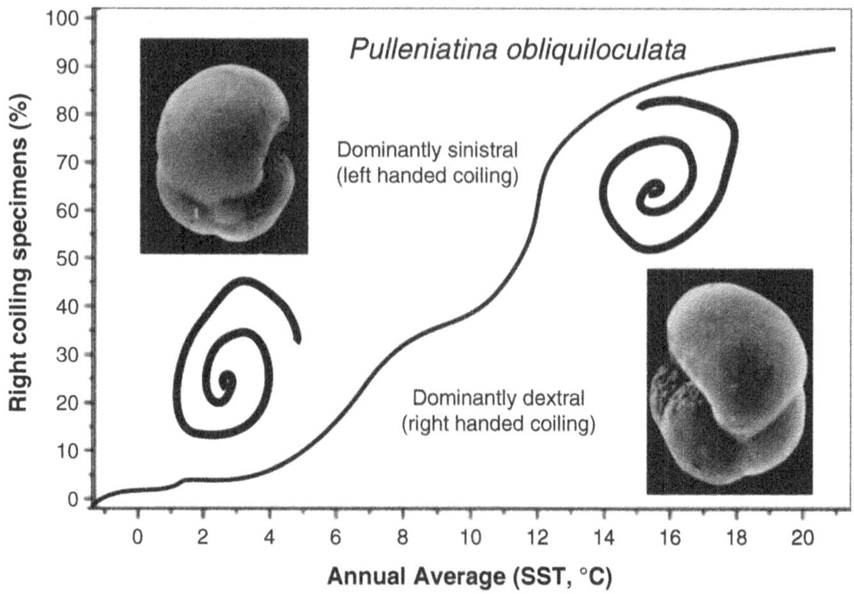


**Fig. 3.12** Changes in coiling direction of shells of the high-latitude species of *Pulleniatina obliloculata* and its relation to SST annual average

the ecological circumstances, a signal measured on a group of specimens may represent the yearly average, in spring, summer, or autumn.

All proxies using species-specific ecology require correct identification of taxa whose behavior does not change through their geographical range. Like physical proxies, geochemical proxies are applied only as far back in time as the relationship between species ecology and biology and the target chemical process remain stationary, because any change in calcification, depth or production season will alter the meaning of the reconstructed record.

At the same time, the huge progress that has been made in the research on planktonic foraminifera over the last years has enlightened a number of basic issues that continue to hamper the use of these organisms for paleoceanographical applications. The nature, origin, and ecological significance of this diversity have to be clarified so that proxies requiring species-specific calibration are adjusted and applied appropriately.

## Age Model

Within a vertical sediment profile (core), measurements of microfossils, and other variables at a range of depths serve as proxy estimates or "proxies" of climate and environmental conditions when the sediment of those depths was deposited. It is crucial to establish reliable relationships between these depths and their ages. Age-depth relationships in marine sediment cores are used to study the evolution of climate/environmental proxies along sediment depth and therefore through time (Lowe and Walker 1997).

Age-depth models for sediment depths containing organic matter, and for ages younger than 50,000 years, radiocarbon dating is used to create an age-depth model. Cores are divided into slices and some of these are radiocarbon dated. A curve is fitted to the radiocarbon data, and interpolated to obtain an age estimate for every depth of the core. The first restriction to be considered is that age should be increasing monotonically with depth, because sediment can never have accumulated backwards in time. As the sediment accumulates through time, changes in accumulation rate will cause fluctuations in the age-depth curve. Many proxies are used to be able to tell time and environment Magnetostratigraphy (another 'science' that deals with magnetic susceptibility of sediments), Nanofossils, radiolarians and Planktonic foraminifera) are illustrate in Fig. 3.13. Often few samples will be dated, and on those few dates, an age-depth model needs to be constructed. Material deeper down in a core will be older than material further up, so, we know that any age-model should increase monotonously, and not show any "leaps back in time".

Another example of integration of calcareous nannofossil, and planktonic foraminifer biostratigraphy is shown on this age-depth plot (Site U1488, IODP WPWP, Rosenthal et al. 2017) where biostratigraphic and magnetostratigraphic horizons indicate recovery of a continuous sequence of pelagic sediment spanning the last ~10 My one of the oceans warmest temperatures occur (Fig. 3.14).

## *Benthic Foraminiferal Species "The Giants of Show and Tell in Space and Time"*

Benthic foraminifera are single-celled eukaryotic organisms similar to amoeboid organisms covered with an organic test that varies from a simple single chamber with an aperture to a complex, multi chambered, perforate, calcareous wall, to an agglomeration of mineral grains embedded in the organic agglutinated test.

Benthic foraminifera are widespread in marine environments, from brackish estuaries to the deep ocean basins and occur at all latitudes. Many species have well defined salinity and temperature preferences making them particularly useful for reconstructing past trends in ocean water salinity and temperature. Many of the taxa have widespread distributions, inhabiting marshes, coastal estuaries, bays, lagoons, and the continental shelf.

They are benthic bottom dwellers characterized by high diversity and abundance where the cytoplasmic body is encased in organic or mineralized test (shell), which provides a fossil record (Cambrian to recent). Due to the presence of the test, benthic foraminifera are useful indicators of past and present-day environmental changes.

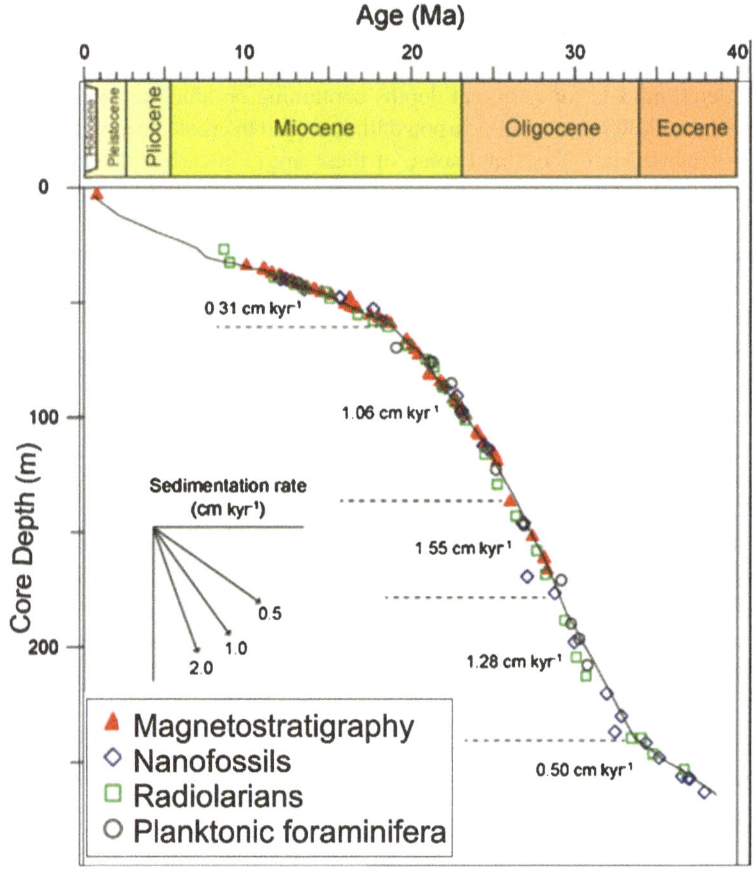


**Fig. 3.13** An age model based on different indicators (Magnetostratigraphy, Nanofossils, radiolarians and Planktonic foraminifera)

Foraminiferal shells consist of single or multiple chambers interconnected with each other by a channel, with only one aperture called *foramen*, which gives the name to the entire Foraminifera group. We will show more about them in this book since they are the "stars of the show" in the microscopic world!! Please see how beautiful these reef species are!! See shells from today and from a million years ago in Fig. 3.15.

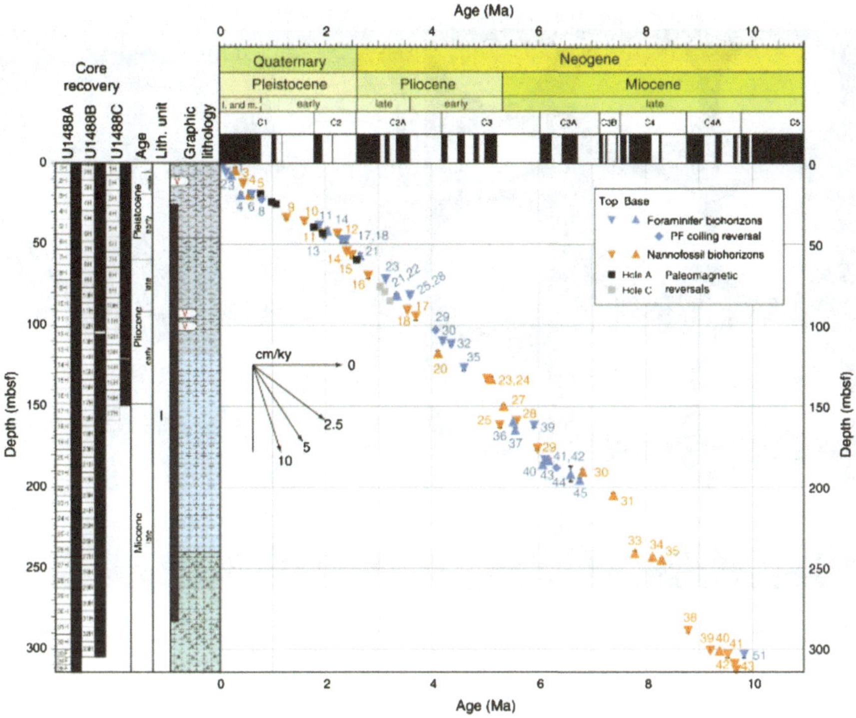


**Fig. 3.14** Age-depth plot for Site U1488. Age of the oldest recovered sediment at Site U1488 is constrained at <10.46 Ma by the absence of planktonic foraminifer (PF) Paragloborotalia mayeri and >9.69 Ma by the presence of calcareous nannofossil Catinaster coalitus at 314.52 mbsf. Sedimentation rates average ~2–3 cm/ky above 50 mbsf and ~4 cm/ky below 50 mbsf (Rosenthal et al. 2017)

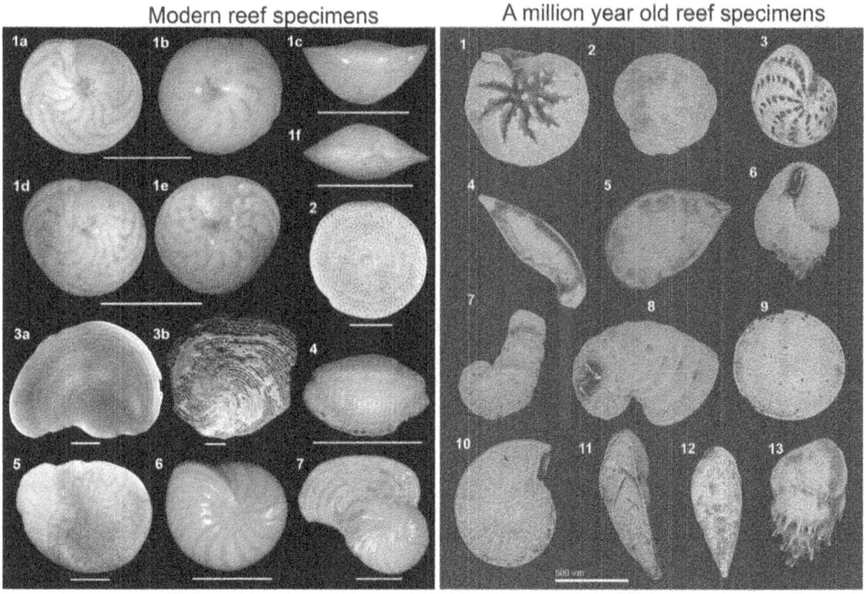


**Fig. 3.15** Modern and paleo specimens from coral reef areas

# Chapter 4
# "Surfers of Bottom Water Masses as Narrators": Water Masses and Geo Properties Interaction as Benthic Habitats for Foraminiferal Bio Indicators

**Abstract** Our "surfers" of bottom water masses and geo properties interaction, the Forams indicators from the marine sediment-water interface dominate unique benthic habitats. In modern oceans, benthic foraminifera proxies are tolerance ranges for salinity, temperature and oxygenation and sedimentological properties and changes in benthic distribution patterns are interpreted as long term changes in bottom-water mass structure around the world. Case studies done in North and South America demonstrate that the "monster" estuarine plume of the Plata River, and the bays in Delaware and Rio de Janeiro are different settings with "subtle" or "abrupt" hydrographic and geological properties changes characteristic of coastal environments. It has also shown that even in these environments subject to seasonal and daily changes, with a tidal movement of the periodic saline water intrusion and freshwater runoff highly dynamic, it is possible to form habitats, all of them accurately occupied and distinguished by Forams as Bio indicators that reflect Freshwater, Cold, and Warm waters, Marine Fronts, Sediment fractions, and dynamics of Salinity and Temperature.

**Keywords** Bio indicators · Quantitative analysis · Long-term changes · Salinity · Temperature · Oxygenation · Water masses

## Introduction

The geographic distribution of live Foram fauna (collected and stained with Rose Bengal) from the marine sediment-water interface is a response to the interaction of bottom water masses and sediment properties, which create unique habitats for benthic indicator species.

To identify water masses in modern oceans, the principal benthic foraminiferal proxies are tolerance ranges for salinity, temperature and oxygen. However, sometimes processes operating at the ocean surface produce effects on the ocean floor,

 
P. P. Beck Eichler, C. P. Barker, *Benthic Foraminiferal Ecology*,
https://doi.org/10.1007/978-3-030-61463-8_4

and can override benthic processes such as seasonality, or input of detritus, which leads to blooms of specific organisms including foraminiferal opportunist species, which is correlated with renewal capacity of marine waters in an aquatic ecosystem.

The development of differentiate thermohaline and oxygenation circulation in bottom waters causes benthic taxa to adjust their depth distribution patterns; some species can appear or disappear, or even be replaced by other tolerant species. Changes in benthic distribution patterns are interpreted as long-term changes in bottom-water spatially and seasonally in many areas worldwide. Besides being influenced by salinity, temperature and oxygen changes, foraminiferal zonation is subject to how sedimentological properties of marine sediments are distributed. This includes how fine or coarse sediment is, and the amount of food available, as organic carbonates react differently in each environment displaying differentiated geo habitats.

The geo habitats for benthic foraminiferal indicator species formed in the recent record are used as a proxy effect of water masses and sediment interaction. Therefore, to narrate our history about bottom water masses and "our foraminiferal bio indicators surfers" of geo properties from benthic habitats, we will show and tell you what will often be in overly long run-on yet very science-y sentences about studies done in North and South America.

The case studies are the "monster" estuarine plume of Plata River influencing the coastal waters of the countries of Argentina, Uruguay, and Brazil; and coastal bays, such as the "wonderful" Rehoboth Bay (Delaware, USA), and the "amazing" Guanabara Bay (Rio de Janeiro, Brazil). These marine environments where georeferenced maps show the beauty of the accurate quantitative analysis when correlating bio indicators and abiotic parameters, unravels distributional and correlation patterns from different settings.

One is as an open marine shelf with "subtle" hydrographic and geological properties, and the other has bays that are enclosed environments where abiotic parameter changes are major and fast. In both environments it is possible to distinguish the bio indicator accuracy revealing the special tool that they are for exposing environmental average conditions and detecting "natural" from anthropogenic induced disturbances, enabling sustainable approaches; namely pointing out where the waters can be renewed so we can explore only what can be replenished, for possible solutions and remediations.

This chapter presents Bio indicators of Benthic Foraminifera that reflect freshwater, cold and warm waters, marine fronts, sediment fractions, and dynamics of salinity and temperature. These conditions are characteristic of coastal environments where balance of tidal movement of periodic saline water intrusion and freshwater runoff establish differentiated possible habitats to be colonized and best renewable and oxygenated sites in both places.

## Case Studies

### The "Monster" Estuarine Plume of Plata River

Freshwater input from Plata River, forms an estuarine plume which flows from South to North in the Southern Hemisphere, and the warm and cold-water masses interaction influences latitudinal distribution of benthic foraminifera along the Southwest Atlantic continental shelf of three countries in South America (Argentina, Uruguay, and Brazil). Factors controlling latitudinal gradients of regional species richness distribution is considered to be variation in the rates of extinction and speciation dispersal history. All of these variations reflect differences in biogeographically or evolutionary processes along latitude, and can be detected by Foraminiferal analysis. Besides the macro-scale distribution, the influence of freshwater runoff and oceanic fronts can disrupt patterns providing interesting phenomena to be explored.

Large-scale Foraminiferal distribution in the South Atlantic was done by Boltovskoy et al. (1980) and it associates the great systems of oceanic circulation, Brazil Current (BC) and Malvinas Current (MC) described by Miranda (1972) between latitudes of 29°S and 35°S. Here we focus on the coastal environments influenced by freshwater runoff from Plata River, and coastal water masses in between these great circulation systems.

Distinct water masses are identified in the study area. The less saline colder Sub Antarctic Shelf Water (SASW) (temperature less than 15 °C and salinity between 33.7 and 34.15), together with the dilution of the "monster plume" in upper shelf waters which takes place at the mouth of Plata River (at 36°S), and at the Patos-Mirim Lagoon (at 32°S) are transported by Malvinas Currents and SASW flowing from South to North. The warmer, Subtropical Shelf Water (STSW) (where temperature is higher than 12 °C and salinity is higher than 34.5) (Piola et al. 2008) are transported by salty and warm ($T > 20$ °C, $S > 36.4$) Brazil Current flowing from North to South (Fig. 4.1).

The low-salinity plume formed by Plata River and the Patos Lagoon outflows form the Coastal Water (CW) that caps the shelf water, leading to a salinity decrease to values lower than 30 (Fig. 4.2). The extent of the low-salinity water has a strong impact on the vertical stratification and in the interface sediment-water as well, and observations and numerical models suggest that the low-salinity estuarine plume extends northward over the shelf penetrating farther north in winter than in summer (Campos et al. 1996; Pimenta et al. 2005; Piola et al. 2005). There is little indication of mixing between these two water masses SASW and STSW, however a formation of an intense temperature, salinity and nutrient front separate these water masses known as Subtropical Shelf Front (STSF) (Fig. 4.1). The front is oriented along the north-south direction, located on average near the 50 m isobath at 32°S and extends southward toward the shelf break near 36°S (Piola et al. 2000).

Salinity, temperature and dissolved oxygenation patterns of bottom waters (winter and summer) in Fig. 4.2 shows the way CW, SASW, and STSW water masses

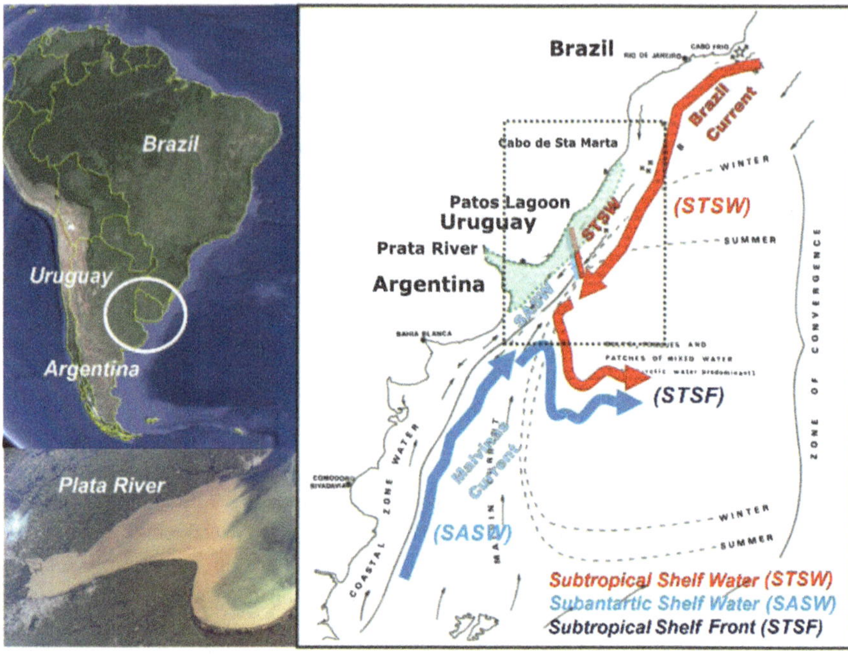


**Fig. 4.1** Location of the study area and the interaction of freshwater runoff from Plata River, and dynamics of coastal water masses (Subtropical Shelf Water, flows from North to South and Sub Antarctic Water, from south to North). (Modified from Boltovskoy et al. 1980). The area of average influence of the "monster plume" of the Plata River during winter is indicated by green shadowing

interact in winter and summer and it also unravels the location of the STSF and its movements, two straight lines in red and blue are the STSF, where the two water masses meet (SASW and STSW). It shows latitudinal gradient of hydrographic distribution varying on space and time.

The clear differentiation of water masses pattern observed in the winter is less apparent in the summer, but the presence of warmer less saline and less oxygenated water characteristic of the water masses and the STSF was detected all over the studied region.

The nature and extent of latitudinal gradients of foraminiferal dynamics and species richness in the oceans are determined by the water masses and the migration (or not) of the STSF and its significant seasonal movement back and forth.

Here we focus on the Foraminiferal indicator species in these water masses interactions describing the complex relations among biological, physical and geochemical variables in the sediment-water interface and how they are responding so they can be selected to be bio indicators.

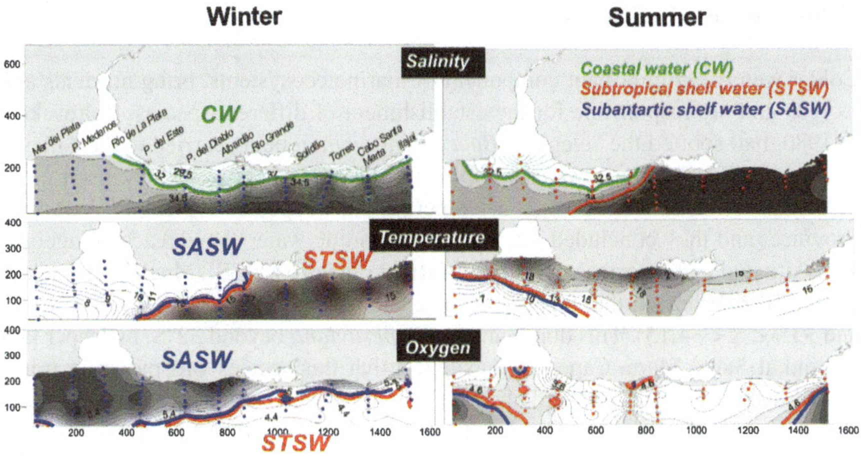


**Fig. 4.2** Latitudinal gradient of salinity, temperature and oxygen varying on space and time showing STSW and SASW, which the STSF is demonstrated by the two straight lines in red and blue showing where two water masses meet (STSW and SASW)

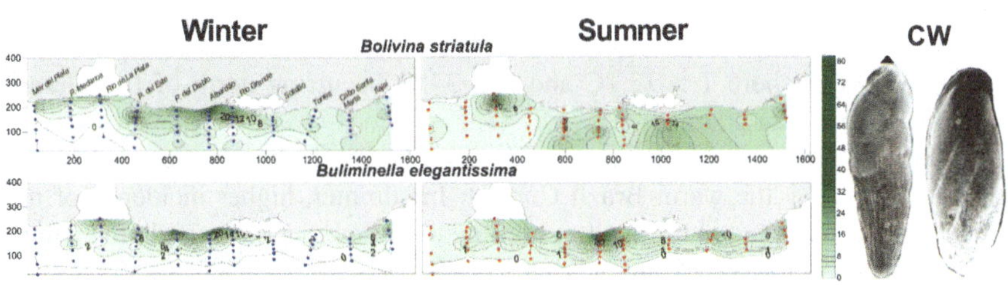


**Fig. 4.3** *Bolivina striatula* and *Buliminella elegantissima* as dominant species of freshwater occurrence in the shelf

## Freshwater Bio Indicators

The Continental Water (CW), a combination of oceanic saline water (SASW) with deriving continental freshwater from continental discharge of Plata River and Patos lagoon, estuarine plume, are detected as far as 50 m depth and presents *Bolivina striatula* and *Buliminella elegantissima* as dominant species (Fig. 4.3).

Boltovskoy et al. (1980) also describe these two species as marine euryhaline tolerant to less saline waters of Patos Lagoon and Plata River. Later on, Eichler et al. (2012b, 2019b) shows them as freshwater indicator of Continental Water (CW) along the Subtropical South American Coast. These examples are being used to illustrate this chapter. This coastal water mass forms low salinity habitats which species of continental origin, estuaries and salt marshes related to fresh water river plumes of Plata River and Patos Lagoon, can survive and thrive in the marine shelf. The ecological pattern is in the marine sediment year-round with seasonal changes, being documented by these wonderful fossils. Both species are from sites were stratification of waters is high, environment is anoxic with high organic matter as well.

## Cold-Water Bio Indicators

Colder waters, an important component of marine ecosystems, bring nutrients and oxygen and are responsible for the establishment of different species. Boltovskoy, in 1980, had spotted the "elegant" *Buccella peruviana* being carried to the Atlantic Ocean via Sub Antarctic waters. Following the same approach, Stevenson et al. (1998) describe this species as a temperate water species typical of Argentine Province, and they concluded that the Sub Antarctic waters have reached lagoonal areas situated in the northernmost part of Rio de Janeiro (22° S). Much later, Eichler et al. (2012b) showed that this species is characteristic of the (SASW) (T < 15 °C and 33.7 < S < 34.15). This dominance of *B. peruviana* beyond 32°S, in deeper layers and at Santa Marta Cape (28°S) suggest that this species was passively transported by prevailing northward shelf currents to lower latitudes in the winter and was succinctly described by Eichler et al. (2008) as is illustrated in Fig. 4.4.

## Warm-Water Bio Indicators

Here we illustrate two species *Globocassidulina subglobosa* and *Uvigerina peregrina* which according to Eichler et al. (2012b) are dominant in the Subtropical Shelf Water (STSW) (where T > 12 °C and S > 34.5) occurring north of Rio Grande (31.5°S) deeper than 20 m.

These two species have been previously described by Boltovskoy et al. (1980) as species carried by the warm Brazil Current. In summer, higher incidence of this water mass is observed in the south according to these bio indicators. In the winter, the south limit of the STSW are the shallowest sites of Torres (28.5°S), and in summer, peak of occurrence of these two species are the deepest sites of Mar del Plata in Uruguay (38°S), suggesting STSW reaches as far south as Mar del Plata seasonally (Fig. 4.5).

## Marine Front Bio Indicators

At 41 m deep, *Bulimina marginata* has a much-defined peak of occurrence in Punta Del Diablo (33°S) within a patch of occurrence between Punta del Este (34.8°S) and Rio Grande (31.5°S) (Fig. 4.6). Eichler et al. (2016), described it as capable of adapting to the high variability dynamics in the transitional area, the STSF, formed

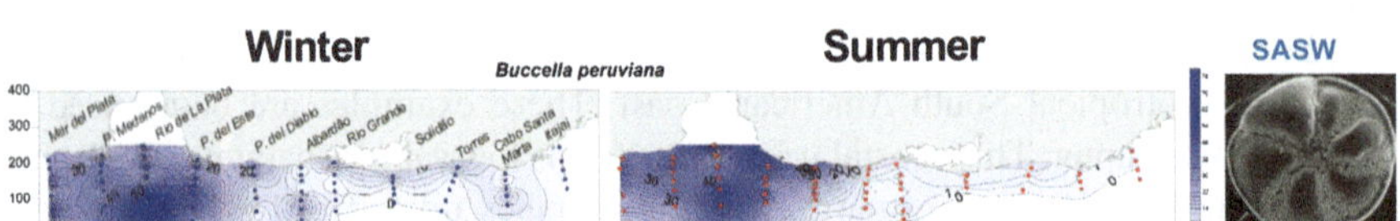


**Fig. 4.4** Dominance of *B. peruviana* in Subantarctic Shelf Water (SASW) as bio indicator of cold water beyond 32°S in deeper layers of waters

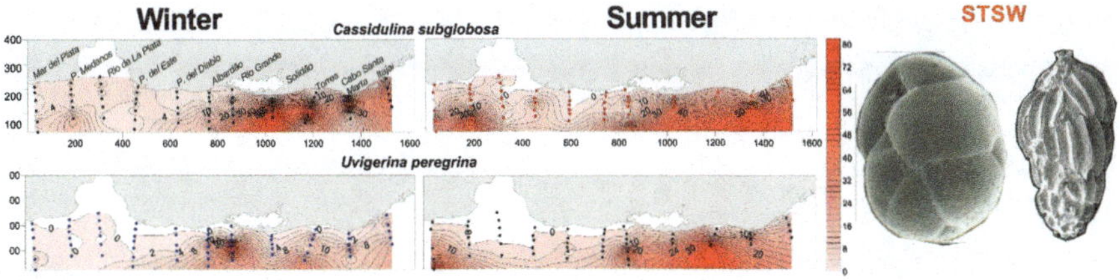


**Fig. 4.5** Dominance of *Globocassidulina subglobosa* and *Uvigerina peregrina* in deeper layers of warm waters of STSW

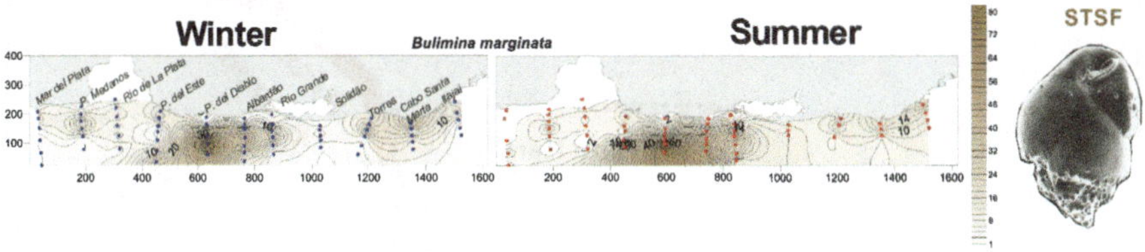


**Fig. 4.6** Peak of occurrence of *Bulimina marginata* where high variability dynamics in the transitional area, the STSF, is formed by the meeting of two coastal water masses (the warm STSW and cold SASW) together with freshwater influence from Patos Lagoon and Plata River

by the meeting of the two coastal water masses (warm STSW and cold SASW) together with a major local freshwater influence from Patos Lagoon and Plata River.

A picture illustrating how easy a marine front is differentiated and a summary of three indicator species and its water masses interaction in the shelf is shown in Fig. 4.7.

## Sediment Fraction Bio Indicators

Spatially distributed *Buliminella elegantissima*, *Elphidium excavatum*, *Pseudononion atlanticum*, and *Bolivina striatula* are bio indicators of sand and mud (Fig. 4.8). *Buliminella elegantissima* and *Elphidium excavatum* occurs where there is less sand, while higher numbers of *Pseudononion atlanticum* and *Bolivina striatula* occurs where mud has lower values on the marine sediment.

## Rehoboth Bay (Delaware, USA) and Guanabara Bay (Rio de Janeiro, Brazil)

The environmental and ecological patterns demonstrate environmental conditions through foraminiferal species dynamics as biological indicators, and the balance of salinity and temperature as physiochemical indicators in bottom waters in these bays (Fig. 4.9).

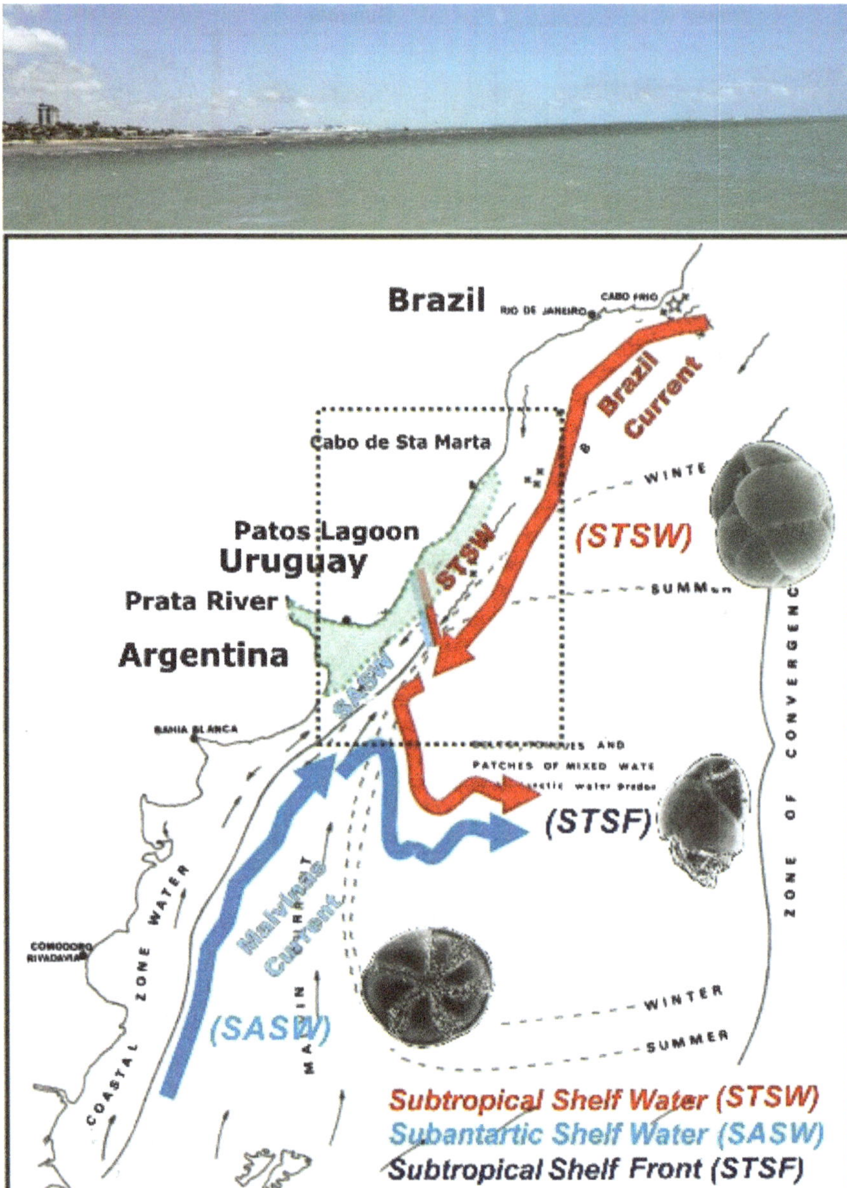


**Fig. 4.7** A picture illustrating an oceanic front, and the summary of three indicator species and its water masses interaction

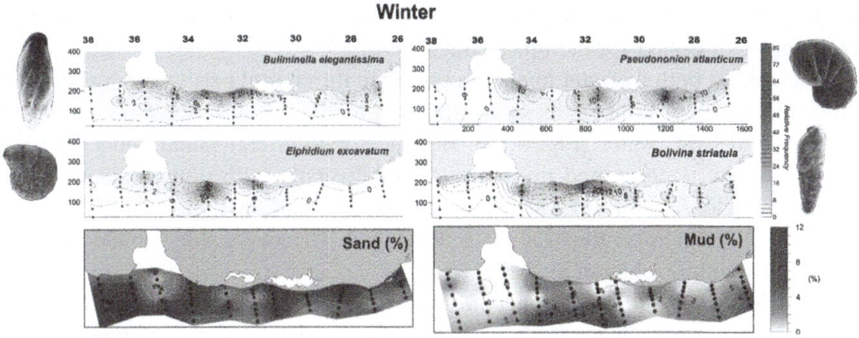


**Fig. 4.8** Bio indicator of sand (*Buliminella elegantissima* and *Elphidium excavatum*) and mud (*Pseudononion atlanticum* and *Bolivina striatula*)

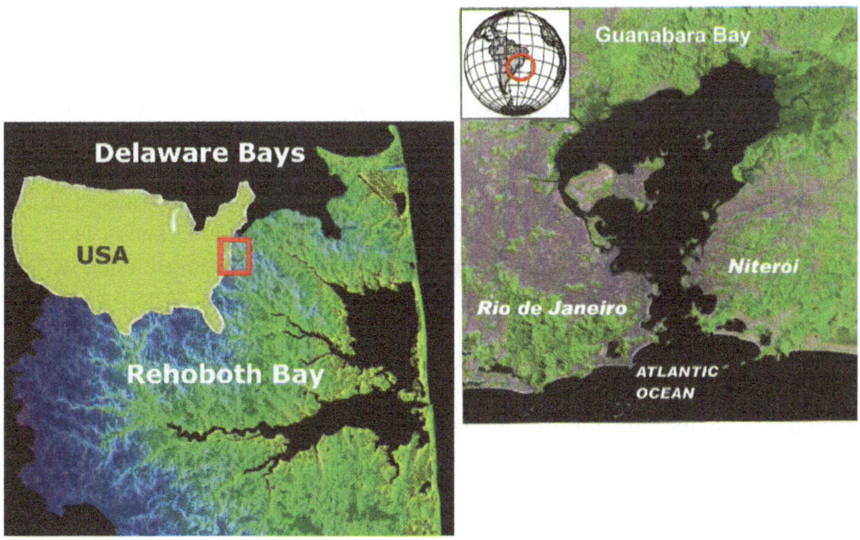


**Fig. 4.9** Rehoboth Bay (Delaware, USA) and Guanabara Bay (Rio de Janeiro, Brazil) and the study of the balance of salinity and temperature in bottom waters

## *Bio Indicator of Dynamics of Salinity and Temperature*

### Rehoboth Bay (Delaware, USA)

Marine waters flow from the south through Indian River inlet, into the Rehoboth Bay, a shallow body of water, part of the Delaware Inland Bays system in southeastern Delaware (USA) (Fig. 4.9) whereas freshwater discharges by streams located in the North and West, by groundwater drainage and by surface runoff from the adjacent watershed. The Inland Bays are characterized by large seasonal variations in surface water temperature (0–27 °C), and surface water salinity (21.8–27) with poorly flushed waters and residence times estimated at 80 days for Rehoboth Bay (Karpas 1978; Center for the Inland Bays 1995).

The coldest bottom-water temperatures are in the central part where it is deepest and bottom-water salinity values range from 27 to 30 in the bay reflecting mixing of marine with fresh water (Fig. 4.10). Benthic foraminiferal fauna collected from Rehoboth Bay is composed primarily of low diversity with opportunistic genera (*Ammonia* and *Elphidium*) tolerating spatial variations in bottom water salinity (19.5–28.7), and temperature (24–28.6 °C) (Fig. 4.10) and are able to trace temperature and salinity (Eichler et al. 2010).

Distributional patterns of *Ammonia tepida* show dominance in the warmest and less saline waters of the central parts of the bay, and *Elphidium poeyanum* inhabits colder and more-saline waters closer to the marine Indian River inlet in the South (Fig. 4.10).

## Guanabara Bay (Rio de Janeiro, Brazil)

During the year, the enclosed bay has lower values of surface and bottom temperatures near its entrance and higher values in the northwest of the bay showing the dynamics of penetration of colder saline water by periodic tidal movements from the adjacent continental shelf and a great influence of the large freshwater contribution (Fig. 4.12). Temperatures in winter ranges from 22 to 25 °C in surface, and from 21 to 23 °C at the bottom. Increase in these values occur in summer, ranging from 26 to 31 °C on the surface and 23 and 29 °C on the bottom, a sharp increase in temperature in the summer for the bay.

The calcareous species *Ammonia tepida* was observed everywhere in the bay (Fig. 4.11), but highest numbers occur in the north and northeast part, close to the mangrove area, the Guapimirim Preservation area where highest bottom and surface temperatures occurs (Fig. 4.11). Highest relative abundances of *Ammonia tepida* occur in the north and northeast regions of the bay, close to the APA of Guapimirim. In these places, the great influence of fresh water from the rivers located in this region, characterizes the region with the highest bottom and surface temperatures and the lowest salinity.

The mangrove area has great influence of fresh waters from many rivers showing lowest salinities in the bottom of the bay, and highest values at the entrance of the bay closer to the Atlantic Ocean. The highest values of salinity in winter range from 31 to 34 in the surface and in bottom salinity, respectively (Fig. 4.12). The increase in fresh water supply from the rainy season (summer) show decrease in this variable, mainly in the bottom salinity close to the mangrove area and rivers. In summer salinity values range from 5 to 32 on the surface and 10 to 34 on the bottom. High values of bottom salinity up until the middle part shows penetration of the saline water in the bottom, similar to the temperature maps explained previously.

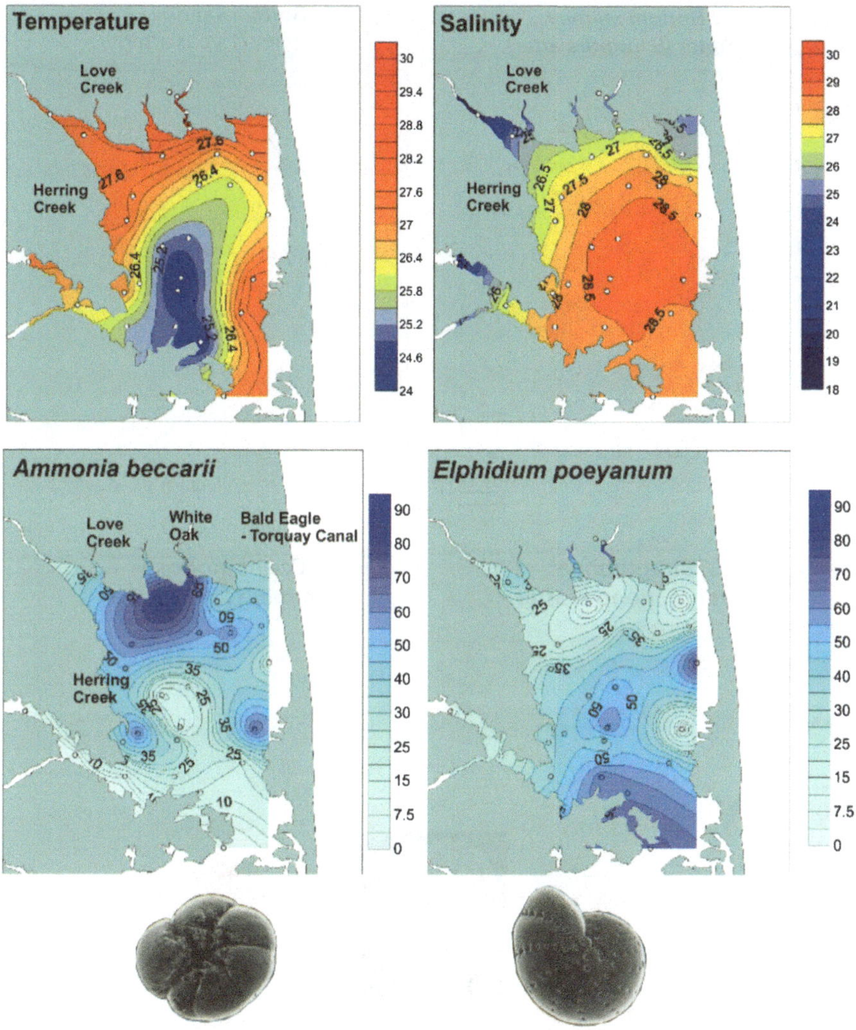


**Fig. 4.10** Temperature and salinity showing a gradient of marine waters flowing from south to the north and west. Relative frequency of *Ammonia beccarii* and *Elphidium poeyanum* distributional patters

Higher abundance of *Cassidulina subglobosa* close to the entrance of the bay (Fig. 4.12) shows its preference for marine environments characteristic and indicates the penetration of cold and more saline waters towards the interior of the bay, similar to the salinity maps.

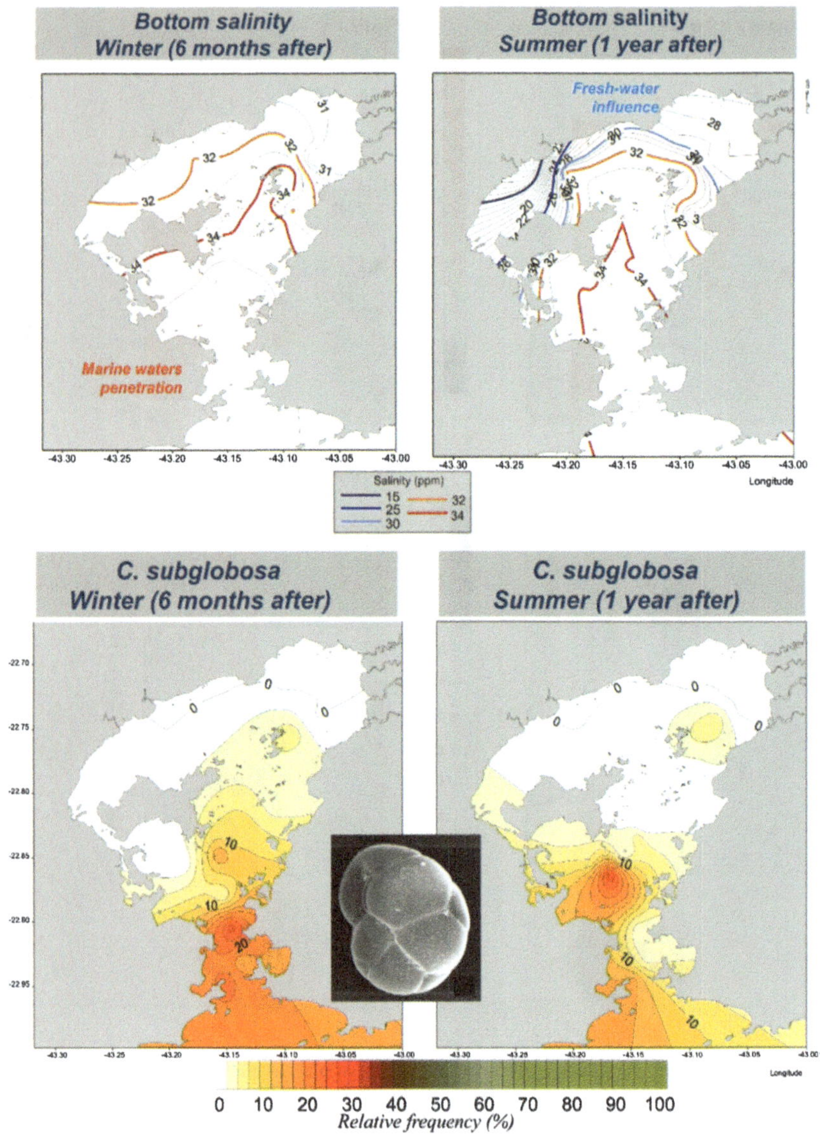


**Fig. 4.12** Bottom salinity and distribution of *Cassidulina subglobosa*

## Silt and Clay Bioindicators

Figure 4.13 shows that silt and clay accumulate in the northeast, north and northwest regions, and sandy sediment at the entrance and in the central parts of the bay. *Bolivina striatula Buliminella elegantissima,* and *Bulimina elongata* characterize environments with lower concentration of silt and clay and are totally absent when

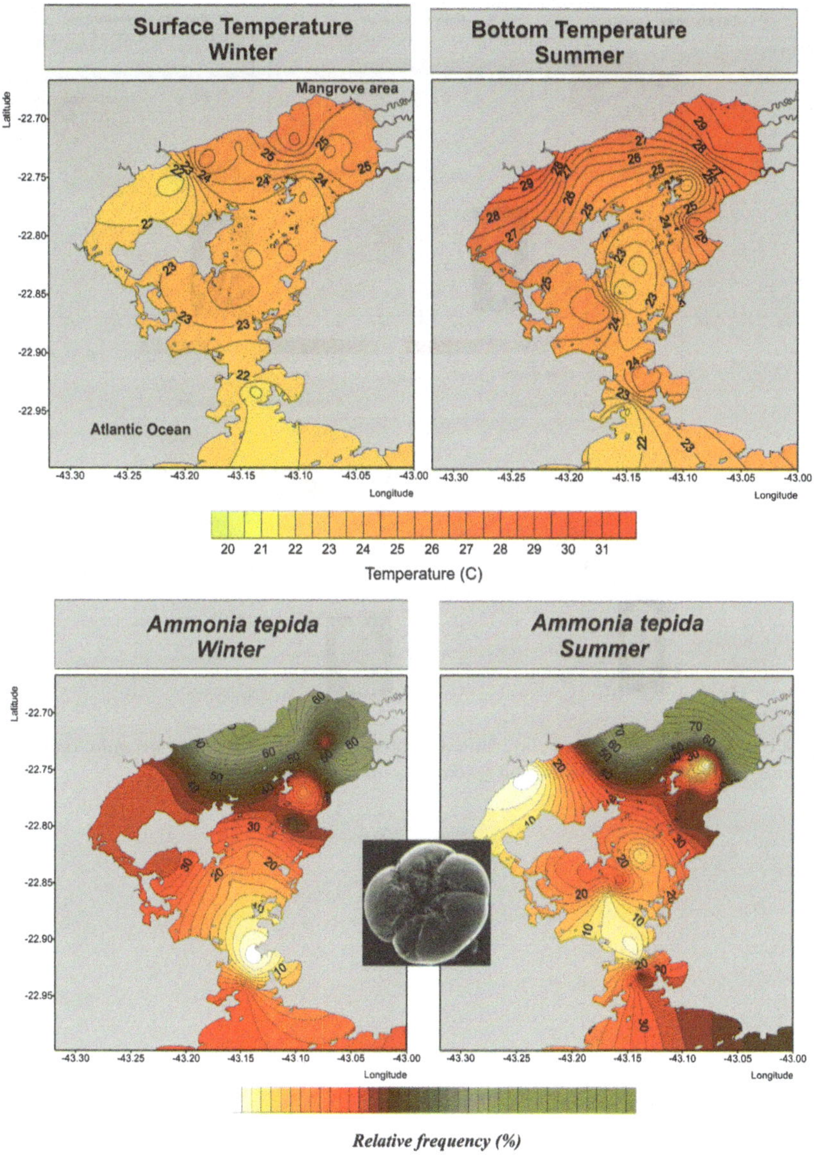


**Fig. 4.11** Surface and bottom temperature, and relative frequency of *Ammonia tepida* distribution

silt and clay values are too high. *B. elongata* prefers the lowest depths, while *B. striatula* appears in the deepest regions. *Quinqueloculina seminulum* by the other hand in general prefers sites where silt and clay have the highest values.

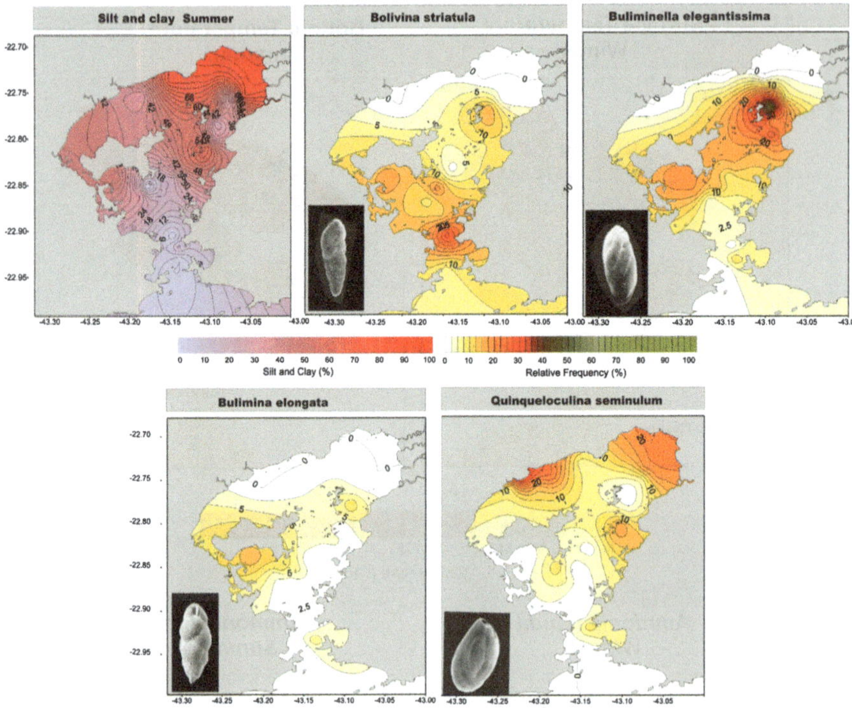


**Fig. 4.13** Percentage of silt and clay, and distributional patterns of *Bolivina striatula, Buliminella elegantissima, Bulimina elongata,* and *Quinqueloculina seminulum*

# Chapter 5
# "Sediment's Quality Responders": Water Renovation, Sewage and Industrial Pollution Stress on Benthic Fauna

**Abstract** Analyses of living foraminiferal and environmental parameters near an outfall at Mar Grosso Beach (Laguna, SC, Brazil) demonstrate its usefulness as indicators of domestic sewage pollution. The low species diversity may be due to sand accumulation in the surf zone of the central part that prevents settlement on or in the substrate. Silt and clay content correlates well with the total coliforms higher diversity of foraminiferal species, reduced salinity, and higher temperatures closer to the mouth of Laguna estuarine system indicate freshwater influence, suggesting relationship between increased diversity and greater availability of terrestrial food, even if the salinity decreases. Higher Forams diversity and average coliform values are closer to the mouth of the estuarine system and under the influence of the outfall. The estuary mouth is under the influence of the outfall and exhibits higher foraminiferal diversity and higher mean values of total coliforms and silt-clay content. Due to the effect of local hydrodynamics, the particulate organic waste derived from the outfall settle down locally, accumulating nearby on the southwestern and northwestern parts of the beach. We confirm our hypothesis that the fine material derived from the outfall is accumulating in shallow parts of the beach. In general, the characteristics of the foraminiferal assemblages show that the environment is in the initial stages of eutrophication. Apparently, the nutrient enrichment of the water by the organic waste has induced changes in the benthic foraminiferal community through an increase in species richness. Due to the high energy of the marine environment, however, the particulate organic waste derived from the outfall does not settle down on the seabed; but it supports the hypothesis that the fine material derived from the 12-m deep outfall is accumulated on the southwest part of the beach.

The fluctuations in salinity and temperature in Todos os Santos Bay established different tolerance limits for certain species, which indicates that species distribution was subject to natural environmental seasonal changes and excellent water renovation. In the present study, many marine species were observed, indicating a strong influence due to intrusion of marine waters from the adjacent continental shelf. Given that, opportunistic-tolerant species directly benefit from certain kinds of contamination of organic substances, or indirectly, through the reduction of competition and predation, increasing their local occurrence; this chapter reveals this bay as an environment where organic matter is easily deposited, the excess of nutrients results in a low diverse fauna dominated by *Ammonia* spp. closer to freshwater sources.



P. P. Beck Eichler, C. P. Barker, *Benthic Foraminiferal Ecology*,
https://doi.org/10.1007/978-3-030-61463-8_5

The occurrence of *Elphidium* spp. and *Quinqueloculina* spp. is limited to the eastern part of the TSB. *Pseudononion atlanticum* and *Bolivina* spp. presented their occurrence limit in the area under continental runoff influence. A decrease in diversity indicates less stability close to Mataripe River and RLAM area, and an increase in diversity in central and east parts of Todos os Santos Bay. Some areas present aliphatic accumulation, and silt and clay accumulation. *Ammonia tepida, Bolivina* spp., and *Fursenkoina pontoni* tolerate sediment with aliphatic and unresolved complex mixture, while *Elphidium* spp. decreases in areas under the influence of pollutants. *Ammonia* spp., *Bolivina* spp., and *Fursenkoina pontoni* are known organic matter and low oxygen-tolerant foraminifers, and in the bay, they are tolerant to petroleum derived pollutants.

**Keywords** Diversity · Dominance · Hydrodynamics · Water renovation · Sewage · Outfall · Salinity · Coliforms · Estuarine system

## Introduction

Scenic beauty and recreational opportunities attract people to the coastal zone which eventually leads to marine pollution. Using modern foraminiferal analyses coupled with oceanographic and geological proxies, we can understand the "water renovation" concept and its applicability in marine pollution.

Distribution of salinity, temperature, diversity, dominance of opportunist species, and organic matter helps us reach an understanding of the ecosystem and its function, showing natural, disturbed, or "optimum state" climax where diversity is so high that it is able to send biodiverse propagules to other less pristine nearby environments increasing their local biodiversity.

In long-term environmental monitoring of ecosystems, the use of benthic organisms as indicators of rate of change in pollution-prone areas is reliable, because the close dependence of benthic organisms on the composition of the marine sediment in which contaminating substances settle and accumulate. Due to slow moving or sessile way of life, benthic communities are used in the monitoring of effects of marine pollution, which render them more susceptibility to local disturbances integrating effects of pollutants over time. They comprise a large number of species, and because of specific sensitivity of species, it is possible to pinpoint subtle impacts reflected in changes in community structure.

In addition, benthos, unlike the plankton or nekton, adapt and reflect local sedimentary conditions over a period of years, instead of just a snapshot of the conditions prevailing at the moment of sediment sampling. Benthic Foraminifera is used in hazard assessment at specific discharge sites, and monitoring of the effectiveness of remedial actions because their short life cycle and quick reaction to environmental shifts. These "fast first-community" responders to disturbing, impacting and distressing agents are used as an early warning of human-induced environmental change, as defined by Kramer and Botterweg (1991).

Variability of salinity, temperature, sediment type, amount of organic carbon, pH, and tidal energy unraveled Benthic Foraminifera as bio indicators of coastal pollution (Boltovskoy et al. 1991; Alve 1991, 1995; Yanko et al. 1999; Scott et al. 2001) revealing that the common effect of increasing concentration of contaminants is reduction in species diversity, until only opportunistic species remain. Such opportunists react favorably to pollution dominating impacted areas of marshes (Schafer et al. 1995; Collins et al. 1995), being tolerant or resistant to the pollutants or contaminants. However, on the other hand, species that are sensitive to specific impact often express that sensitivity through their absence (Yanko et al. 1999). The foraminiferal assemblages impose themselves as a means of assessing the quality of waters and aquatic ecosystems, for the biological variables, associated with salinity, temperature, dissolved oxygen content, pH, sediment characteristics, and heavy metal concentrations provide data indicative of depositional and erosional patterns, pollution, contamination, and eutrophication levels as well.

Distribution patterns of benthic Forams species in shallow, enclosed marine environments, contaminated by organic and industrial wastes have demonstrate that they are sensitive and inexpensive biomarkers capable of indicating sediment quality and water renovation potential.

This chapter talks about these findings illustrating two examples in Brazil, one in the south, and other one located in the northeast. One is a Beach in Mar Grosso Beach, downtown Laguna (Santa Catarina State), where this beach is under the influence of a domestic submarine outflow discharge, and the other one, the Todos os Santos Bay, Bahia state is influenced by refinery wastes. This chapter intend to enlighten the reader in the wonders of the hydrodynamics capacity of increase local biodiversity, and we are here to explain how we are able to say these things.

## *The Popular Mar Grosso Beach!!*

Here we show how foraminiferal species dynamics, and total coliform bacteria, responds to physiochemical (salinity, temperature, dissolved oxygen, pH) and to geological proxies (percentage of sand, silt, and clay) helping us to evaluate the ecological conditions resulting from the outfall discharge in the sampling setting of Fig. 5.1. Summer sampling was carried out on the Mar Grosso Beach, a very important tourist destination in the south of Brazil, close to where the effluents from the submarine outfall are released.

Sampling was done on the beach adjacent to the lagoon's entrance where the submarine outfall is located (Mar Grosso Beach). The area receives contribution from freshwater and marine water masses, being highly dynamic with coarse grains dominating in most parts, and some exceptions where silt and clay predominate.

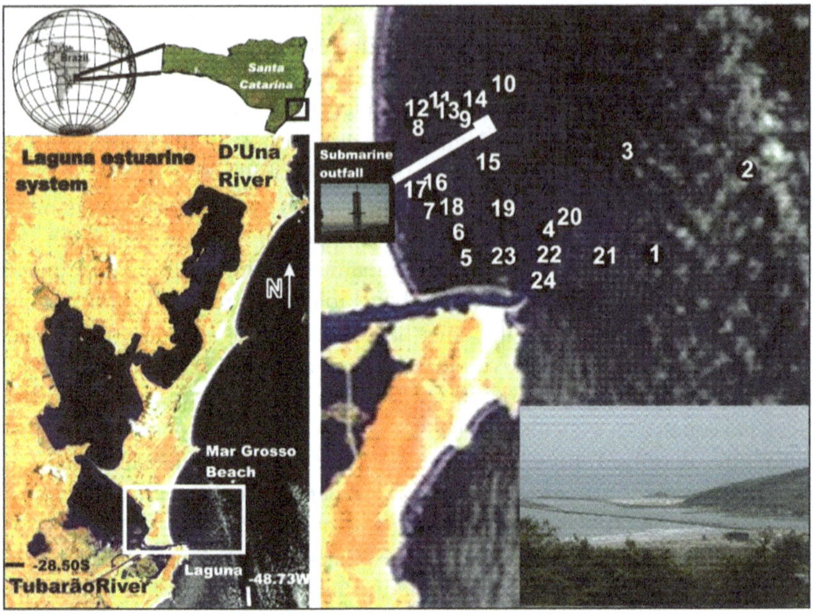


**Fig. 5.1** Sampling location of the lagoons. View of the submarine outfall and the lagoons' estuarine mouth, the only communication with the sea

Our hypothesis is that the fine material derived from the outfall is accumulating on the southwestern and northwestern parts of the beach.

## *Water Renovation and Domestic Sewage*

All over the world, domestic sewage is an inevitable consequence of human settlement and activity, the coastal zone is affected by alterations introduced to the water, and it has been increasing in recent decades. This growth of coastal urban centers have resulted in an increase of the amount of nutrients and other harmful material in the sewage, including organisms, which are not welcome at all, such as pathogens. This continued growth calls for the establishment of adequate strategies to manage, reduce, or eliminate the impact that sewage wastes have on the environment and on human health. The water renovation is the probability that the ocean provides a cleaning effect on the waters, and especially bottom waters, giving the environment the capacity to auto regenerate.

The restrict environments like bays are prone to be totally polluted or contaminated and it is directly related to the ocean intrusion in the environment. However, open sea can also be contaminated by sewage being transported by wind, surface circulation, or along the ocean bed from polluted coastal zones to other zones (Castro and Miranda 1998). Bottom waters bathe marine sediments and therefore benthic habitat has the ability to tell the quality of conditions. A healthy bottom habitat is necessary to maintain the organisms that live in the water column as a whole. Clean water with potential periodic natural renovation is of "high ecological status" and, successful environmental implementation and management requires a thorough understanding of ecosystem structure and functioning under pre-impacted conditions in order to set realistic restoration goals for the problems of sewage and industrial pollution challenges like sediment settlement of urban wastes in specific situations.

Since Forams geo habitats are fossilized and allow tracking of periodically natural and environmental changes due to human impacts over the last hundreds of years, ecological studies employ different multi-variate statistical techniques for interpreting modern foraminiferal environment relationships. Benthic foraminiferal communities living in marine and coastal environments respond to a broad range of biotic and abiotic factors that result in complex interactions in space and time. The first case of this chapter deals with quantitative analyses of living foraminiferal as indicators of domestic sewage and environmental parameters near a submarine outfall at Mar Grosso Beach (Laguna, Santa Catarina, Brazil).

In this stretch of Mar Grosso beach where the submarine outfall is situated, we evaluated the response of the foraminiferal microfauna and fecal coliforms to physical, geological and chemical gradients, and point out the area most affected by the sewage plume. Mar Grosso beach, a humid temperate, with an average annual temperature of 19.7 °C is situated in Tubarão city region, about 118 km south from Florianópolis, the Santa Catarina State Capital. This coastline of 208.706 km is subjected to micro tides, with the wind playing an important role on the ocean dynamics. The outflow of this estuarine system is in the southeastern part of Mar Grosso Beach.

## *The Submarine Outfall: The Eww of Sewage Making the Gross of Mar Grosso Beach!!*

The domestic sewage on the coast of Laguna discharges effluent residues and wastes into the open sea from a submarine outfall, about 1400 m long, at a depth of about 12 m off the Mar Grosso beach, one of the most popular beaches in the southern region of Brazil.

The effluent shows obvious increases in summer, causing biological, chemical, physical and geological alterations in the marine environment, because the introduced fresh water and excessive nutrients disturbs the basic conditions of the ecosystem thus interfering with the ecological chain. Salinity changes affect the water-mass stratification; nekton and benthos, the consumers, take advantage of the extraordinary increase in food availability (Eichler et al. 2003).

Besides Eichler et al. (2012a) samplings which are illustrating part of this chapter, no other studies or assessments of ecological consequences and fauna responses due to the installation (in 1986) of the submarine sewage outfall, and the extent of its influence on the sediment quality of the Mar Grosso beach and surroundings have been done. Will this gross negligence and subsequent contamination continue to accumulate indefinitely? There are impact questions needing immediate address that remain unanswered.

Domestic sewage directly released into the sea is a problem worldwide, whether in terms of the volume of substances or in terms of practical problems of public health. Pathogenic organisms contained in fecal material transmit diseases specially this new coronavirus, so we should also consider the potential of water renovation from nature to help fight pandemics. Nitrogen and organic phosphorus from human fecal material, together with other organics that flow into the sea and can cause diseases, are normally recycled into their inorganic forms. If this amount of material is too much, it will promote an increase of planktonic algae (eutrophication), which in turn will drastically reduce dissolved oxygen, causing fish death from low oxygen. Other impacts, such as the accumulation of sand, silt, clay, and organic material, and a reduction of water transparency from suspended particulate matter, produce negative visual effects. These features will worsen as time passes and become unbearable for local and touristic alike.

## *Hydrographical Parameters*

Increasing from West to East, depth of sampling varied between 4 and 20 m, with shallowest areas in the south part of the beach. Higher salinity in the eastern and southeastern sectors of the study area, and lower at the mouth of the estuarine system and near the outfall. It is possible to see zonation of this variable, where lower salinity occurs in the south most part. Higher temperatures occur near the mouth of the estuarine system. Oxygen content remains constant above 7 mg/l, pH did not show significant variation, and the lowest value occurred in between the outfall and the mouth of the estuarine system. High salinity and oxygen peaks with low temperature occurred in the southwestern part of the beach. Figure 5.2 illustrates these conditions.

## *Geological Parameters*

Sediment pH shows lower values in two different areas of the beach. One is closer to the outfall and the other one is in the central part of the beach. Higher amounts of silt and clay were noted in the southern sector adjacent to the mouth of the estuarine system, and in the northwestern sector. The distribution of silt and clay shows the influence of the plume from the outfall and the Laguna estuarine system, suggesting that clay does not settle in the central sector because of the turbulence generated by the confluence of continental waters into the ocean.

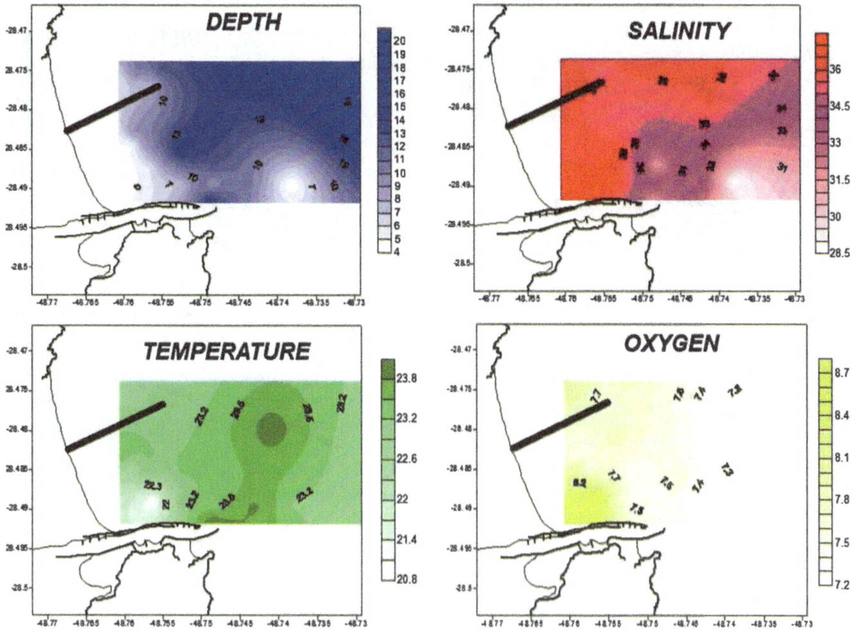


**Fig. 5.2** Depth of sampling, salinity, temperature, and oxygen content at the beach

Accumulation of "natural" origin due to the influence of the estuarine system occurs rather adjacent to its mouth, whereas in the southeastern and northwestern sectors its provenance seems to be from the effluents discharged by the submarine outfall and thus of anthropogenic origin. No accumulation of silt or clay in the central sector, however, there is a sand bank formation starting at the entrance of the outfall, in the western sector of the beach. The sand accumulation seems to have occurred because the circulation generated by the balance between northeasterly and southerly winds. Sand, however, is being accumulated at the outfall. Due to the effect of local hydrodynamics, the particulate organic waste derived from the outfall does settle down locally accumulating nearby the southern part of the beach and the entrance of the outfall (Fig. 5.3).

## *Biological Parameters*

Small numbers of living foraminiferal species in the area, and some sterile samples with no organisms at all. These barren ones were located in the surf zone, where strong water dynamics hinder growth of foraminiferal communities. Fecal coliforms were absent at the majority of sampling stations, however, two stations located in the southwest part of the beach were exceptions, having high coliform abundance. These sites where silt and clay also accumulate, seemed to be one of the

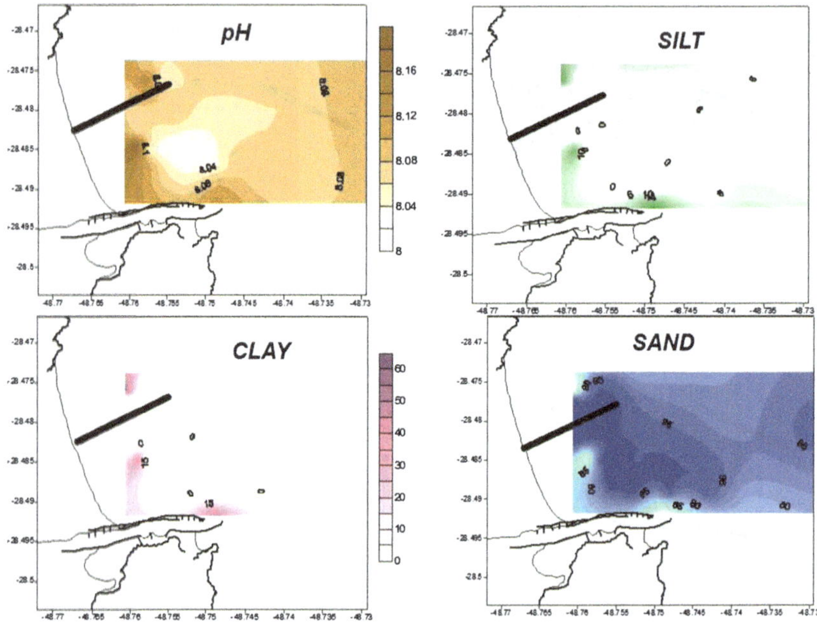


**Fig. 5.3** Sediment pH, amount of silt, clay, and sand at Mar Grosso beach

worst sites in regards to the health of the ecosystem. As for total coliforms, we note that the samples with very high values of total coliforms and those with high clay and silt content were barren of Foraminifera, and low concentrations of total coliforms were found related to the higher salinity.

The spatial distributions of dominant foraminiferal species *Pseudononion atlanticum, Elphidium* spp., *Buccella peruviana,* and total coliforms are shown in Fig. 5.4.

*Buccella peruviana* had reduced abundance at sites where the values for total coliforms were higher. Higher values of coliforms are closer to the mouth of the estuarine system and under the influence of the outfall. The abundance peaks of *Pseudononion atlanticum* and *Elphidium* spp. were at stations with highest water temperatures. The fecal coliform group consists of a variety of bacteria, including the genera *Klebsiella, Escherichia, Serratiaa, Erwenia,* and *Enterobacteria.* The use of fecal coliform bacteria as indicators of sanitary pollution is regarded as more significant than the use of "total" coliform bacteria, because fecal bacteria are restricted to the intestinal tract of warm-blooded animals. Overall, the concentration of coliforms is an important parameter that may indicate the existence of pathogenic organisms, including those responsible for the transmission of water-borne diseases such as typhoid, fever, paratyphoid fever, bacillary dysentery, cholera, and proliferation of already mentioned coronavirus. A sterile zone for Foraminifera linked to a lack of silt and clay and consequently of organic material and food for the species, is present with low numbers of total coliforms.

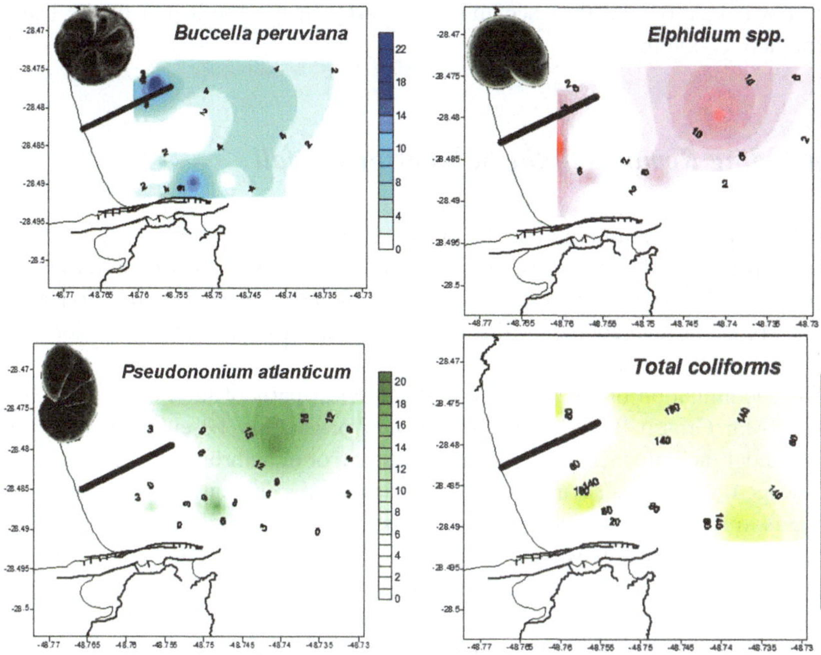


**Fig. 5.4** Distribution of *Buccella peruviana, Elphidium* spp., *Pseudononion atlanticum,* and total coliforms

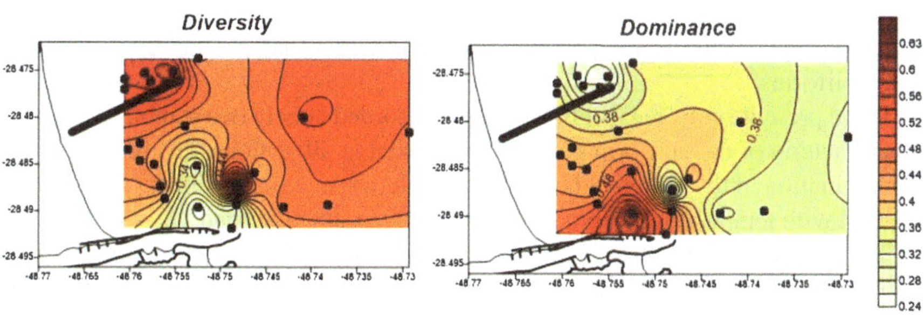


**Fig. 5.5** Diversity and dominance of foraminiferal fauna in the beach

A plume of lower diversity follows until the middle of the beach being more pronounced closer to the mouth of the estuarine system. Higher diversity is observed in the submarine outfall. Higher values of foraminiferal dominance closer to the mouth of the estuarine system and under the influence of the outfall the dominance is lower (Fig. 5.5).

Based on foraminiferal diversity and average coliform count, higher values are closer to the mouth of the estuarine system, and under the influence of the outfall. There is accumulation of fine particles of silt and clay and high correlation with total coliforms in the southern part of the beach. In addition, due to the effect of local hydrodynamics, the particulate organic waste derived from the outfall does settle down locally, and thus accumulates nearby in the south part of the beach where there is dominance.

## *Concluding Remarks of the Submarine Outfall*

Low diversity of foraminifera and the presence of barren zone near the outfall may be partly related to the surf zone, where hydrodynamics hinders benthic foraminiferal colonization. However, the sand accumulation and lack of clay and silt deposition in this sector severely reduce the food supply, and also prevent the development of a foraminiferal community.

The accumulation of silt and clay in the deepest parts of the inner continental shelf of Mar Grosso Beach suggests that their origin may be either of "natural" origin, under the influence of the estuarine system, or of anthropogenic origin, from the sewage outfall. This accumulation of fine-grained sediment has promoted local increases in foraminiferal diversity. The deposition of total coliforms in the central part of the beach probably was facilitated by a circulation associated with the northeasterly and southerly winds that prevail in the region.

Higher dissolved oxygen all over was not a limiting factor that negatively impact any foraminiferal species, on the other hand, the temperature increase observed at the mouth of the Laguna estuarine system favored the proliferation of *Elphidium* spp. and *Pseudononion atlanticum*. The distribution of *Buccela peruviana,* also representative of other foraminifera species (*Ammonia tepida, Bulimina marginata, Buliminella elegantissima Elphidium* spp., *Orbulina* spp.) are responding to temperature, depth, salinity, and percentage of sand rather than percentage of clay and total coliforms.

In the present study, fecal coliforms were recorded at only two stations. However, other bacteria of the coliform group were present at all stations, demonstrating the contamination of the region although foraminifera species are showing a better correlation with temperature, depth, salinity, and percentage of sand.

The continuing increase of the sand bank would reduce water depth and may impede water circulation, especially in the deeper parts of the southwestern and northwestern sectors of the Mar Grosso beach; these sectors are currently seriously affected by silt and clay accumulation, and oceanic bottom waters are not capable of renewing the system. The foraminifera benthic fauna has undergone degradation in the central sector, because the accumulation of sand and the absence of fine material here are conducive to the development of a barren zone, preventing the proper flush of bottom waters.

Populations of coliform bacteria, however, exist all over; their development does not seem to be related to the amount of silt and clay in the sediment. The accumulation of silt and clay in the study area is linked to both a "natural" source, influenced by the estuarine system, and an anthropogenic source, i.e., the sewage outfall. The deposition of material of both origins is affected by water circulation associated with northeasterly and southerly winds, and the low species diversity may be due to sand accumulation in the central part. By the other hand, higher diversity closer to the mouth of Laguna estuarine system, where reduced salinity and higher temperatures indicate freshwater influence, suggests a relationship between increased diversity and greater availability of terrestrial food. At present, fine particles such as silt and clay do not accumulate in the immediate neighborhood of the Mar Grosso submarine outfall, but the deposition of sand in the central near the mouth of the estuarine system may cause a stagnation problem in the future. There is accumulation of fine particles of silt and clay and high correlation with total coliforms in the southern part of the beach that are not being exported, being also a problem concern for the near future.

## Todos os Santos Bay, the Oil Refineries in Bahia!!

For 2 consecutive years, from 2003 to 2005, the composition of Benthic Foraminifera organisms' assemblages was used to assess the occurrence of different bio facies and to establish pollution and contamination bio indicators of the environmental quality of marine sediments in the ecologically differentiate environments of Todos os Santos Bay. This region holds the Landulpho Alves Refinery (RLAM) area from PETROBRAS, and this study shows the influence of the industrial pollution on the benthic communities using a hypoxia index.

## Species Distribution

No anomalies in foraminiferal tests were detected. Year round, mangrove areas harbor less diversity areas than sites closer to the refinery. In summer there are more fragmented tests were than in winter. *Ammonia* spp. dominates in the region, followed by species characteristic of marine environments under little influence from continental runoff such as *Elphidium* spp., *Pseudononion atlanticum* spp. and *Quinqueloculina* spp. The presence of agglutinated forms (*Gaudryina exilis, Ammotium* spp., *Haplophragmoides wilberti, Arenoparrella mexicana, Trochammina* sp.) was also found, indicating high influence from continental runoff in some stations.

## Hypoxia Index

To determine the degree of environmental pollution of marine sediment through the hypoxia (A-E index) we applied the index described by Sen Gupta et al. (1996), and it application was based on the absolute abundance of two common species (*Elphidium excavatum* and *E. poyeanum*). Species of *Ammonia tepida* are tolerant to hypoxia, although species of *Elphidium* are less resistant than those of *Ammonia*.

The A-E index is calculated by the formula:

$$\frac{A.B.Ammonia}{A.B.Ammonia + A.B.Elphidium} \times 100$$

where: A.B. = absolute abundance (number of individuals/50 $cm^3$).

Percentage of organic carbon and concentration of PAH values were correlated to A-E index and *Ammonia tepida* in plots, and the analysis of tests fragmentation were observed in relation to the state of test preservation, due to acidity and transportation.

Positive correlation between distribution of *Ammonia* spp. and PAH (polycyclic aromatic hydrocarbons) concentration in two seasons, and the A-E index was also correlated to organic carbon, and increases in the hypoxia index and in organic carbon content (Fig. 5.6).

## *Concluding Remarks Regarding the Industrial Pollution*

The study of foraminifera in the bay shows occurrence of marine species in environments under low freshwater runoff influence such as *Ammonia* spp., *Elphidium* spp., *Pseudononion atlanticum, Buliminella elegantissima* and *Quinqueloculina* spp., and of agglutinated forms. This kind of fauna in the Todos os Santos Bay (TSB) indicates that marine waters dominate the environment, however low diversity and strong dominance of a few species is characteristic of estuarine environments where

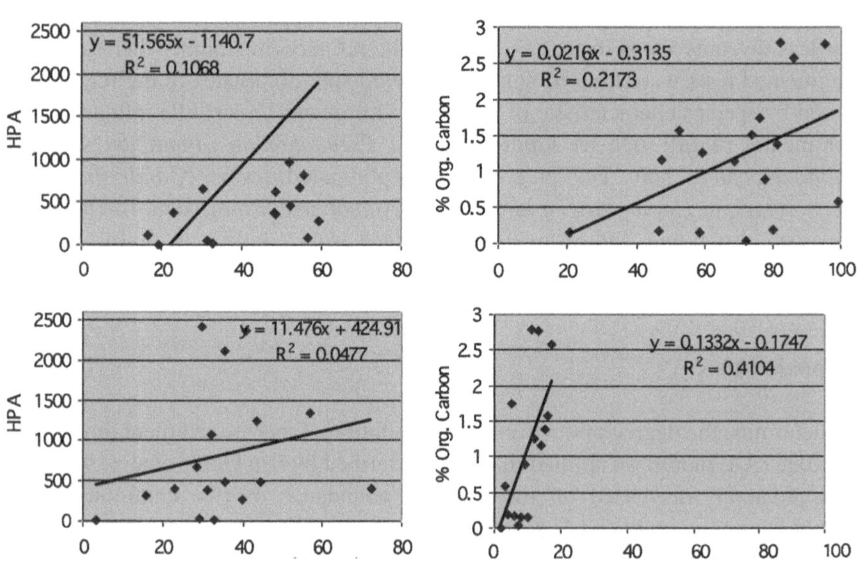


**Fig. 5.6** Correlation between distribution of *Ammonia* spp. and PAH (polycyclic aromatic hydrocarbons) concentration, A-E index, organic carbon content

fresh water limits organism distribution, or yet of not very well preserved environments.

The opportunistic species *Buliminella elegantissima,* known for its high tolerance to high organic matter content (Setty and Nigam 1982; Eichler et al. 1995; Sen Gupta 1999), had same distribution pattern throughout the year. Occurrence peaks for this species were found especially where PAH levels were high. The absence of *B. elegantissima* and of *Pseudononion atlanticum* in some stations suggests high influence of continental runoff in winter and summer. These estuarine species occur more associated with strong marine influence. The dominance of *Ammonia* spp., especially at stations where the highest organic phosphorus, PAH, UCM and aliphatic levels were found, reaffirms its status as an opportunistic species. On the other hand, *Elphidium* spp. dominated where a decrease in PAH, UCM and aliphatic was observed.

According to some authors, *Bolivina* spp is a species typical of environments rich in organic matter, which is mainly retained in muddy sediments, found practically in all regions of TSB. Environments with high organic contents have a direct relation to high mud concentrations (Tyson 1995), since in this type of sediment energy is quite low, contributing to the processes of organic matter deposition. In addition, fine sediments have a tendency to retain heavy metals and other pollutants (Kjerfve et al. 1997). Figure 5.7 shows that dominant species found in Todos os Santos Bay are *Ammonia tepida, Elphidium* spp., *Bolivina* spp., *Pseudononion atlanticum* and *Quinqueloculina* spp.

The number of species per sample increases when the environmental features are more typically marine (Phleger 1970). It was thus to be expected that this bay presented a higher diversity per sample level than more restricted estuarine environments. Although this bay is under the influence of continental runoff in the summer and winter, the foraminiferal assemblages indicate that this influence is quite small.

Comparison of the present study with data from less saline estuarine environments shows that number of species found is similar to this bay. This fact demonstrates that the pollution effects superimpose themselves onto the natural

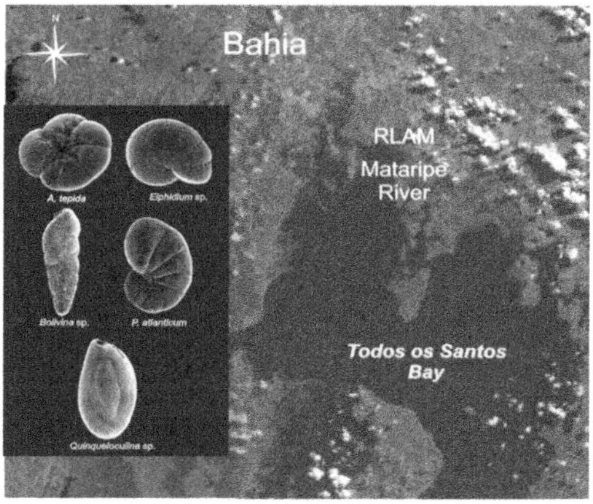


**Fig. 5.7** Dominant species found in Todos os Santos Bay: *Ammonia tepida, Elphidium* spp., *Bolivina* spp., *Pseudononion atlanticum* and *Quinqueloculina* spp.

environmental factors, limiting the establishment of non-opportunistic species and facilitating the development of the opportunistic ones. Those species that are abundant in polluted areas are obviously tolerant to the pollutants to which they are subjected, whereas other species can manifest their sensitivity to the same pollutants by their absence (Yanko et al. 1994).

The accumulation of pollutants in coastal areas does not rely entirely on the supply of these materials by rivers, but also on the chemical interaction between those elements and the sediment constituents, and are therefore reflected by the carbon and fine sediment contents, which more easily adsorb these types of particles (Baptista-Neto et al. 2000).

Analysis of relative abundances suggests that the species *Gaudryina exillis, Textularia earlandi, Arenoparrella mexicana, Haplophragmoides wilberti, Trochammina* sp., *Ammotium* spp., *Ammoastuta inepta,* and *Miliammina fusca,* which occur in places more subjected to continental runoff, were grouped. The marine species *Globocassidulina subglobosa* and *Uvigerina* spp. are species rarely found in the study region, since they are usually associated to colder water masses. In the summer the influence of marine species was observed by the presence of *Discorbis* spp. Cardoso (2000), studying São Sebastião Channel (São Paulo State, Brazil), and Eichler et al. (2001), studying Bertioga Channel (São Paulo State, Brazil), suggest that the occurrence of this species was related to high-density marine currents penetrating via the bottom. Therefore, in the south region of the RLAM subtidal area, the penetration of the marine current was evidenced by the presence of *Discorbis* in that region, showing the potential to renew the water system of the bay.

High-diversity peaks were observed in the central and eastern regions of the RLAM subtidal area, while smaller values were found in the northern region close to the mouth of the Mataripe River and immediately offshore from the RLAM. That region displays a high dominance of *Ammonia* spp. Evenness is distributed along a gradient, where the smallest environmental stability values occur close to the RLAM and the Mataripe River continental runoff, indicating less environmental stability in those regions, and increase toward the region more subjected to ocean influence to the south. As environmental stress increases, species diversity falls, resulting in an increase of dominance (Odum 1988).

The north-to-south diversity increase probably reflects local hydrodynamic, where marine with high salinity waters are responsible for water renovation in the southern portion of the bay. The data obtained in the present study reveal that the northern sector, including the regions close to the RLAM, are low-circulation zones, and are thus less subject to water renovation. The analysis carried out over 2 consecutive years shows a relative environmental stability in terms of foraminifera. The higher diversity figures found in the second monitoring year reflected the inclusion of stations with more marked euhaline characteristics in the RLAM subtidal area and therefore do not have an apparent relation to eventual inter annual changes.

# Chapter 6
# "Climate Change Tiny Giants": Ocean Warming and Acidification Changes: Response of Microfossils of the Anoxic Great Seas

**Abstract** Our results show that the severely organically enriched benthic environment of Flamengo Inlet and Florianópolis Bay (São Paulo and Santa Catarina states) have been recently invaded by the agglutinated foraminifera *Trochammina hadai.* The invasion in Flamengo Inlet occurred sometime before 2010, and by 2014–2015, the species has become one of the dominant faunal constituents at several locations. This species probably arrived in this region due the active commercial fishing and recreational boating traffic. Some questions remain unanswered and we still have to investigate exactly how it invaded into this embayment, and at what other locations in Brazil the species has been introduced. We do know that in Florianopolis North Bay, in the interface marine sediment-water, extremely low dissolved-oxygen values at the bottom waters associated with high percentage of total organic carbon, and low pH (high acidity) are better tolerated, by fauna of opportunists (*Ammonia tepida* and *Buliminella elegantissima*) and *Trochammina hadai,* than by other calcareous benthic foraminiferal species. Abundance of organic matter, low oxygen, with low pH are optimum for the fauna replacement and establishment of the invasive *T. hadai,* thereby enhancing its chances of proliferating. The invasive species is replacing calcareous benthic foraminifera, especially *Elphidium* genera, but needs further studies to find out if there are other genera being damaged as well. On the other hand species that thrive in high oxygen, that were found closer to the marine entrance towards the north of both environments, are responsible for bringing constant periodic natural oceanic water, renewing to both environments. We conclude that periods of hypoxia is unfavorable for bottom infauna where just the opportunists predominate in some central sites of both marine ecosystems severely affected at least by seasonal hypoxia.

**Keywords** Acidity · Carbon dioxide · Temperature · Oxygen · Florianopolis Bay · Flamengo inlet · Influences

 
P. P. Beck Eichler, C. P. Barker, *Benthic Foraminiferal Ecology*,
https://doi.org/10.1007/978-3-030-61463-8_6

## Introduction

Besides sea level rise, the main consequences of climate change are ocean warming and its evil twin acidification. Ocean acidification has been identified as 'a future global climate change impact concern' because it has been slowly affecting entire ecosystems, and it is a threat to local economies, especially shellfish and fisheries production. We warn that rather than a future concern, it is a 'right-now concern—and a major future impact threat'. The scientific-community-at-large has limited understanding of ocean acidification impacts, yet researchers continue to monitor and evaluate local conditions in many parts of the world. Marine bay environments are some of the richest and most biodiverse areas in the world. Ocean current circulation and upwelling of deep cold waters brings nutrient rich waters to the surface at certain times of the year, increasing productivity and oxygen. Bays act as a carbon sink. Global oceans absorb 25% of our carbon dioxide emissions. Therefore, when there is an excess of carbon dioxide emissions in the atmosphere, this excess is absorbed in seawater and in the marine sediment interstitial water, causing a series of chemical reactions that lowers pH, increasing ocean acidity. These "ocean acidification" changes have serious implications for our coastal ecosystems, altering marine life behavior and development—yet this remains largely unaddressed. Increased acidity interferes with the process by which calcifying organisms such as crab, oysters, mussels, and certain plankton and benthic Foraminifera build their shells. Foraminifera are the base of the ocean food chain as first consumers in the ocean. If anything adverse happens to them, it is evidenced with immediacy, and the entire ecological chain suffers and declines. Here we show how coastal ecosystems directly respond to anthropogenic threats.

### *The "Tiny Giants" as Environmental Indicators*

Because ancient Forams shells are microfossils and their ecological dynamics remain well preserved in ocean sediments and continental sedimentary rocks for over millions of years, their fossilized shells evidence the environmental conditions when the organism was alive. They are therefore called "Environmental Indicators" because they allow the study of the ecological and oceanographic present conditions, back to past ocean conditions.

They act like clocks, since foraminifera can be used to evidence the age of sediments and rocks that contain them. They are used as thermometers, because their shell chemistry is used to reconstruct climate history of the ocean. In addition, they are also indicators of sea level rise, because these conditions alter environmental properties in the interface sediment-water and Foraminiferal communities respond precisely to it.

This is possible since each species thrives in a particular environment having a range of preferable seawater conditions (nutrient and oxygen content, temperature, water depth, water mass, ocean currents). When the ecological chain is compromised, other species like crab, mussels, and rockfish, and entire fishery communities that play a huge role in coastal economies, are also in danger. It is projected that 30% economic revenue loss will occur in the next 50 years due to ocean acidification. This singularly important multiple billion-dollar revenue loss issue alone, is one that merits motivation immediacy toward thorough research, and the necessary funding regarding this subject.

Unfortunately, nowadays bays are usually regions impacted by chronic and acute environmental stress caused by anthropogenic influence, such as domestic sewage, oil and gasoline leaks, and erosion, among many other factors since they receive all effluents including human waste. These results increase organic matter, showing that enclosed and restricted embayment is strongly affected by organic pollution. This is particularly true for bays over the last century having suffered severe environmental degradation without any efforts taken to measure the damage or prevent it. In marine sediments of polluted bays, high organic matter is acidic, with low oxygen and pH. This condition favors Forams agglutinated tests rather than calcareous ones because of dissolution processes.

Another environmental variable of ecological importance to estuarine and coastal marine ecosystems around the world, which has so drastically changed in a very short period, is the dissolved oxygen content. The use of benthic foraminifera associated to the study of bottom waters is an accurate alternative for interdisciplinary environmental studies in the areas of oceanography and environmental monitoring, because foraminifers are able to synthesize the dominant characteristics of the environment very effectively, highlighting short and long-term natural changes reacting to natural variables and to anthropic effects. They are unicellular, possessing either a carbonate or an agglutinated shell, which is left in the sediment as a fossil. After the animal dies These animals are found throughout the marine habitat and the majority thrive in highly oxygenated environments, however some species tolerate low levels of oxygen (hypoxia) and even anoxia (no oxygen).

Seasonal anoxic or hypoxic conditions have been observed in temperate coastal environments and in tropical marine environments. The frequency, severity, and length of hypoxic conditions are mainly a consequence of a hydrographic structure causing water stagnation, or by the inflow of deep oxygen-depleted waters from adjacent regions.

The decrease of dissolved oxygen in bottom waters, together with a stable halocline or thermocline, induces high periods of oxygen stress that may have severe ecological consequences on bottom fauna and demersal fish. While hypoxic and anoxic environments are natural and have occurred through geological time, their occurrence in shallow coastal and estuarine areas now appears to be increasing rapidly, most likely accelerated by human activities. Residues of natural and synthetic substances derived from anthropogenic activities are important vectors of the degradation of coastal environments. The oil industry, heavy maritime traffic and domestic effluent sources pose risks to marine environments. Industrial and domestic wastes contribute to the eutrophication of this environment, causing damage to marine fauna and flora. Such damage affects the trophic chain leading to ecological imbalance. Reaction from benthic forams fauna tracks these consequences.

Historical data on several large ecosystems from the turn of the nineteenth century shows that hypoxia was never reported in the "Kattegat Sea" (marine waters between Sweden and Denmark), and nowadays experiences severe seasonal hypoxia. Synthesis of literature on benthic hypoxia and anoxia reveals that the oxygen budgets of many major coastal ecosystems have been affected mainly through the process of eutrophication, which is the production of excess organic matter.

It appears that many ecosystems are now severely stressed by hypoxia, nearing or at the threshold of change or collapse (loss of fisheries, loss of biodiversity, and alteration of the food chain). However, dissolved oxygen does not appear to be the limiting factor for some opportunist species that live in the upper most one centimeter of sediment (Matsushita and Kitazato 1990), so oxygen-depleted conditions deeper within the sediment should not have a dramatic impact on the abundance of low oxygen tolerant species. These assemblages are associated with high input of anthropogenic organic matter and thrive when conditions are less than optimum.

These Climate Change "tiny giants" indicators reveal present and past anoxic, oxygenated, acid, or warm oceans conditions. Since each species thrives and responds in a particular environment having a range of preferable oceanographic conditions of seawater and sediments such as organic matter, nutrient, oxygen content, temperature, water depth, water mass, and ocean currents. Such natural features and/or environmental problem patterns are enhanced and the opportunist species dominates, proliferates, invades, or even replaces original native species.

## *Bays Worldwide and Invasive Species*

Increases of opportunist agglutinated foraminiferal species *Trochammina hadai* (hereafter *T. hadai)* correlate positively with input of organic matter, nitrogen, carbon, and sulfur in Osaka, Japan (Tsujimoto et al. 2006a, b, 2008), in Gamak Bay, South Korea (Lee et al. 2012), and in a mollusk facility in the inner bay of Wando, in South Korea (Lee et al. 2016). They are also linked to eutrophicated bays. *T. hadai* has also been reported dominating foraminiferal assemblages in many other contaminated brackish water locations. In Japan, assemblages with >50% *T. hadai* have been observed in the stressed environments of Akkeshi Bay (Morishima and Chiji 1951), Matsushima Bay (Matoba 1970), and Hamana Lake (Ikeya 1977), whereas core top sediment samples from polluted Yokohama Port yielded 100% *T. hadai* (Toyoda and Kitazato 1995). These assemblages were assumed to be associated with high input of anthropogenic organic matter.

*T. hadai* was identified as invasive species from Japan in estuarine sediments of San Francisco Bay (SF Bay), California, United States by McGann and Sloan (1996, 1999). Only later McGann (2014a, b) realized that this species first appeared in SF Bay as early as 1983 (only 1.5% of the fauna), and that by 2000 the species was in

96% of samples within the bay (McGann et al. 2000). The proliferation of *T. hadai* in SF Bay is associated with a decline in relative abundance of the common native foraminiferal calcareous *E. excavatum* influencing the ecological dynamics of native foraminiferal fauna because of fauna replacement.

We are still unsure what effect the invasion of the agglutinated species have on biodiversity of native fauna in Bays of Brazil. Documenting the occurrence of this invasive microorganism and the impact of its introduction to the local food chain is important and worthwhile to spread awareness, create interest, and help others become conscious of the deleterious problems in the ecological chain. In addition, mechanisms involved in the species invasion are still uncertain. At this point we think that *T. hadai* probably arrived in the bays of Brazil with commercial fishing and recreational boating. However, we still do not understand what other mechanisms are responsible for its introduction, all of the bays this species has invaded, and which of those are potential sources for ecological chain acidification. Although we know that they are linked because usually bays are close to large cities and therefore subject to many identified and non-identified sources having collateral eutrophication and acidification effects.

Events of acidification, dissolution and fauna replacement with *T. hadai* have occurred in Brazil in two organically enriched benthic environments of eutrophicated bays in São Paulo and Santa Catarina States, and this will be illustrated here in this chapter. One of the bays is in the southwest of Brazil named "Saco da Ribeira" in the Flamengo Inlet, northern coast of Ubatuba (São Paulo), and the other in the south named "North Bay of Florianópolis" (Santa Catarina).

## *"Saco da Ribeira"*

São Paulo State (Brazil), a severely organically-enriched benthic eutrophicated bay environment has been invaded by this opportunistic species *T. hadai.* The invasion described by Eichler et al. (2018) in Saco da Ribeira occurred sometime before 2010, and by 2014–2015 this species became one of the most dominant faunal constituents at several locations in this Bay. The arrival of *T. hadai* coincides with sharp reduction in the abundance of *Elphidium excavatum,* illustrating some of this chapter.

Flamengo Inlet, covers an area of about 18 $km^2$ with an average width of 2.5 km, open to the sea in the south, is relatively shallow (maximum depth about 14 m), and harbors a confined area called "Saco do Ribeira" (Fig. 6.1). Foraminiferal distribution from marine sediment surface samples provided the opportunity to document the replacement of calcareous *Elphidium excavatum* by *T. hadai* for the first time in this small yet very popular harbor with many boats in the "Saco da Ribeira" Flamengo Inlet, northern coast of Ubatuba, São Paulo, Brazil.

Bottom temperatures here ranged from 23.7 to 28.9 °C (mean of 25.7 °C), with colder values occurring at the southern part closer to the sea, and warmer values in the north and west. Highest value of bottom dissolved oxygen of 5.48 mL $L^{-1}$ was

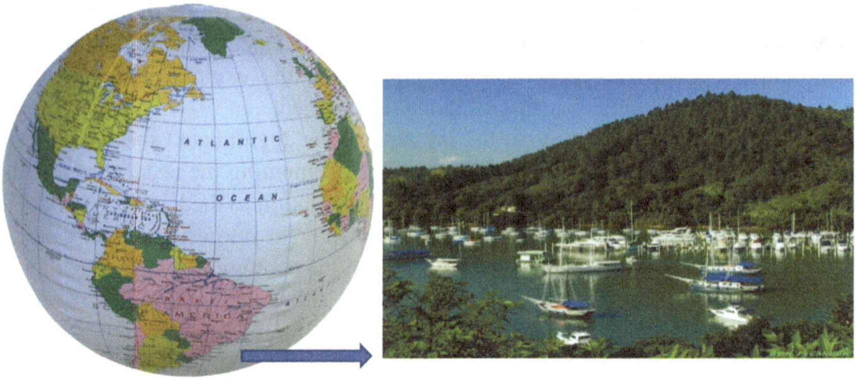


Fig. 6.1 Saco do Ribeira, an enclosed bay, located on the northern coast of São Paulo State, Brazil

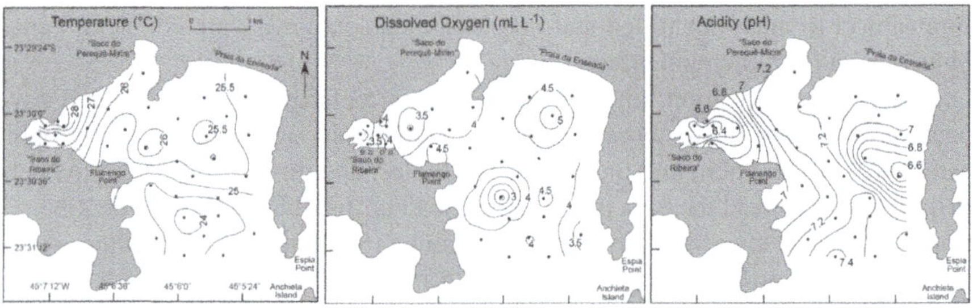


Fig. 6.2 Pattern distribution of bottom temperature, dissolved oxygen, and pH in Saco do Ribeira

in the northeast, and lower values of 3.77 mL $L^{-1}$, 2.87 mL $L^{-1}$, with the lowest value of 1.76 mL $L^{-1}$ in the central part. Bottom water acidity (pH) followed a similar distribution pattern, with the highest value (7.45) in the southern most oceanic site, and the lowest value of 6.28 in two locations within "Saco do Ribeira" (Fig. 6.2).

These values are the result of processes both of natural eutrophication due to the seasonal advance and retreat of South Atlantic Central Water (SACW) and contributions associated with ongoing human activities (Rodrigues et al. 2014). Phytoplankton blooms occur in Flamengo Inlet, resulting from an enrichment of the euphotic layer due to low energy, stratified hydrodynamics, and deposition of fine sediment with high organic carbon content, especially in the inner parts of the inlet (Magliocca and Kutner 1965). These blooms support the establishment of a benthic fauna in coarse to fine silts, the finest of which (i.e., clayey silt) are found in the inner portion of "Saco do Ribeira" (Mahiques et al. 1995; Rodrigues et al. 2014).

Ocean surface pH levels are predicted to fall by 0.3–0.4 pH units by the end of the century, and coincide with an increase in sea surface temperature of 2–4 °C. The combined effect of ocean acidification and warming on the functional properties of foraminiferal shells is largely unknown and is of growing concern as the shell provides protection from mechanical and environmental challenges. After exposure to acidification, significantly reduced foraminiferal shell thickness occurs in many species due to dissolution. We show that under near-future climate change conditions, ocean warming and acidification will pose a greater risk to foraminiferal communities even if their food availability is unlimited.

Hypoxic conditions in the Ubatuba region was registered as early as 1994 with no mention of agglutinated foraminifera. Later in 1998 and 1999, samples taken in Ubatuba Bay recovered agglutinated specimens *Ammobaculites, Ammotium and Gaudryina, but no T. hadai* (Burone and Pires-Vanin 2006) and the very tolerant agglutinated species *Miliammina fusca* (Eichler et al. 2004). Rodrigues et al. (2014) recovered the agglutinated *Ammotium, Gaudryina, Glomospira, Haplophragmoides, Reophax,* and *Textularia,* as well as five species of *Trochammina.* They also did not report the presence of *T. hadai*. However, by 2014–2015, this invasive species had been recovered at 78% of all samples in the northern portion of the inlet from "Saco do Ribeira" to "Praia da Enseada", comprising a significant portion of the assemblage (=13%).

The exact mechanism by which *T.* hadai was introduced into Flamengo Inlet is not known yet, however, we do know that, release of ballast water from transoceanic ships and bulk carriers is a vector for introduction of invasive macro and microorganisms in ports and estuaries according to Carlton (1985); Williams et al. (1988); Carlton and Geller (1993); Smith et al. (1996); Chu et al. (1997); Gollasch et al. (2000, 2002); Dobbs et al. (2003); and Radziejewska et al. (2006).

Saco da Ribeira is a small port that is known for commercial fishing and tourism. Here, more plausible vectors of introduction would be anchors and anchor chains (Carlton et al. 1995) on fishing and recreational vessels acquiring sediment (primarily mud) from neighboring ports harboring *T.* hadai, or from fishing bait from other ports (Lau 1995; Cohen et al. 2001) similarly carrying *T. hadai.* The invasion of *T. hadai* in Flamengo Inlet is a new opportunistic colonization, in a region already impacted by chronic and acute environmental stress caused by anthropogenic influence, such as domestic sewage, oil and gasoline leaks, and erosion among others (Rodrigues et al. 2014). Organic matter values up to 11% in the sediment is excessive, suggesting that the embayment is very strongly affected by organic pollution. This is particularly true of "Saco do Ribeira" where, over the last century, severe environmental degradation has occurred, without measures taken to prevent it. The total organic matter averages 7.4% here, and surface sediments are acidic, with two localities having a pH of 6.28 (Rodrigues et al. 2014). The latter condition favors agglutinated tests such as those of *T. hadai* instead of calcareous tests.

In contrast, dissolved oxygen does not appear to be a limiting factor in "Saco do Ribeira" despite the fact that very low concentrations of dissolved oxygen have been reported within the sediment, which is insufficient for many species to live in, resulting in their migration to surface layers to find oxygen. *T. hadai*, however, lives in the uppermost one centimeter of sediment (Matsushita and Kitazato 1990) and has been observed living at the sediment-water interface, so oxygen-depleted conditions deeper within the sediment should not have a dramatic impact on its abundance.

Clearly, the environmental conditions in Saco da Ribeira Bay, with extremely high percent total organic carbon and high acidity in some areas, are better tolerated by *T. hadai* than by other foraminiferal species, thereby enhancing its chances of proliferating.

We still do not know what effect the invasion of the agglutinated species *T. hadai* will have on the biodiversity and native fauna of this Brazilian harbor, however we know that in the state of Alaska a Regional Citizens' Advisory Council officially lists *T. hadai* as species of immediate concern as invasive (Prince William Sound Regional Citizens' Advisory Council 2004).

For these reasons, it is worthwhile to continue documenting the spread of this invasive microorganism in Flamengo Inlet and the impact of its introduction on the local food chain. It is also of interest to determine what mechanisms were responsible for its introduction, what other Brazilian harbors the species has invaded, and which of those were potential sources for Flamengo Inlet.

*T. hadai* probably arrived in the bay due to the active commercial fishing and recreational boating traffic, but we still have to understand what other mechanisms are responsible for its introduction, and in what other bays has this species has invaded. In addition, which of those were potential sources for ecological chain acidification should be further investigated. Usually bays closest to large cities are subjects of many identified and non-identified sources having the collateral effect of eutrophication and acidification.

Environmental conditions in the interface sediment-water of marine sediments in Bays, with extremely high percentage of total organic carbon and high acidity because of low pH in some areas, are more likely tolerated better by an agglutinated Foraminifera species, such as *T. hadai,* rather than by other calcareous benthic Foraminiferal species, thereby enhancing its chances of proliferating. When other calcium carbonate benthic Foraminifera endemic fauna is dissolved by acidity in the sediment, there is a possibility of the establishment of opportunistic—not dissolvable—Foraminifera fauna. Agglutinated opportunist species are made of small sediment grains and therefore are not subject to acidification. They do not make their own calcium carbonate shells, instead they look for either different or same size grains, depending if they are selective or not they agglutinate same size grans or different sized ones.

Figure 6.3 shows two different shell compositions. Figure 1a presents the agglutinated *T. hadai* and 1b shows the calcareous *Elphidium excavatum* shell.

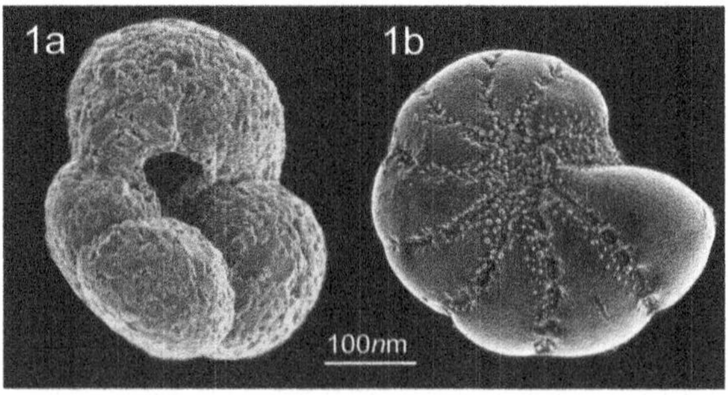


**Fig. 6.3** (**a**, **b**) Two different shell composition. 1a presents *T. hadai* and 1b shows *Elphidium excavatum*

## North Bay of Florianopolis

Along the North Bay of Florianopolis, several events of oxygen depletion have been occurring, and fish mortality has been observed but never studied. A periodic decrease of fauna has occurred, mainly in soft-bottom mollusks. Some unpublished reports briefly explained such mass mortalities by abnormal increases in water temperature. Preliminary studies in this Bay have identified an abnormal decrease of dissolved oxygen in some sites at two different situations. It appears that many ecosystems that are now severely stressed by hypoxia may be near or at a threshold of collapse (loss of fisheries, loss of biodiversity, and alteration of food webs).

To describe this process of anoxia, we have monitored water quality variables, the physical chemical structure of the water column and the foraminiferal fauna associated in the benthic substrate analyzed for either natural or anthropogenic influences, as causes of the environmental disturbances in this bay in two distinct periods, autumn and spring. We intend to evaluate seasonal hypoxia and investigate the dynamics and quality of the North Bay of Florianopolis and the ecological health of bottom sediment through environmental indicators in this stressed environment that is already under pressure from tourism and urban activities.

Here we also want to document the occurrence of the estuarine agglutinated foraminiferal invasive species *T. hadai* in the low oxygen habitat of marine sediments analyzed. This species was first described suggesting that influences in ecological dynamics of native foraminiferal fauna under replacement in acidic environment is yet to be determined. With few studies about fauna replacement, in dyoxic microhabitats in Brazil, we still do not know how the species dynamics is and how they compete.

Florianópolis is the capital and second largest city of the state of Santa Catarina, in the South region of Brazil. The city encompasses Santa Catarina Island and surrounding small islands, as well as part of the mainland. The North Bay is limited by the Santa Catarina Island to the East, and by the mainland to the West (Fig. 6.4).

The bay includes many mangrove areas and has recently become a major sea farming site providing important income for a community of oysters and mussels growers. Jellyfish are often a nuisance for both human leisure activities and shellfish cultivation. The last few years have seen growing public awareness and an increase in reported sting events, leading to growing media publicity. Florianópolis has occasional episodes of fish mortality and explosions of medusas but have no published records. Currently, unprecedented population of macro medusa explosions are being recorded there and worldwide, which may be an indication that major changes in marine ecosystems are taking place in relation to eutrophication and dissolved oxygen in the bottom waters and marine sediment.

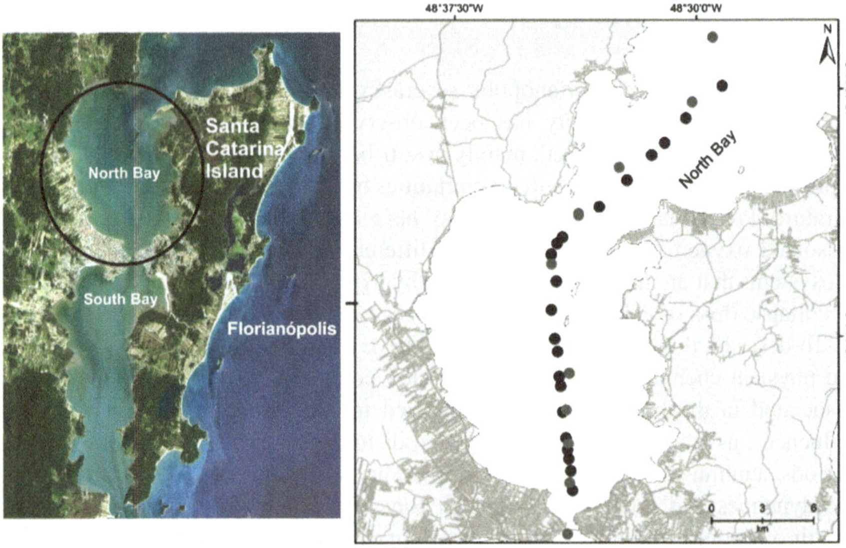


**Fig. 6.4** Study area and sampling stations

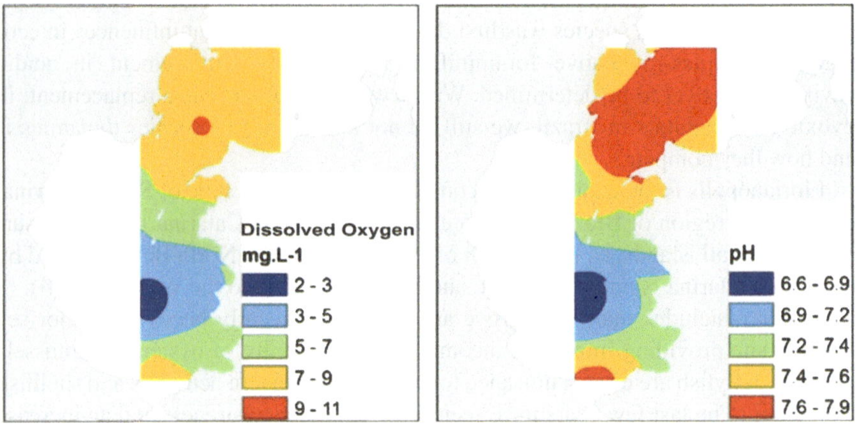


**Fig. 6.5** Dissolved oxygen for bottom waters, and pH for marine sediment

## *Water Dissolved Oxygen and Sediment pH*

Bottom oxygen range from 4.7 to 6.5 mL $L^{-1}$ and sediment pH varies from 6.8 to 8 (Fig. 6.5). Central parts have less oxygenated waters with lower ph. In some sites of the central parts, bottom dissolved oxygen values decreased to reach conditions of hypoxia (1–2.6 mg $L^{-1}$) or anoxia in springtime.

## *Biological Parameters*

The dominant living species are *Buliminella elegantissima* (35%), *Ammonia tepida* (26%), *Bolivina striatula* (7%), *Elphidium* sp. (6%) and *Pseudononion atlanticum* (6%). The first three are present in all samples tolerating low oxygen rates and high concentrations of organic matter. *Buliminella elegantissima, Ammonia tepida, Bolivina striatula* are opportunistic species and can tolerate changes that other species cannot resist, such as high concentrations of heavy metals, sewage and household waste. *Elphidium* sp. (6%) and *Pseudononion atlanticum* (6%) prefers more oxygenated and saline waters showing that there is a strong marine influence.

Some samples evidenced the presence of the invasive species, the agglutinated *T. hadai* also replacing existing fauna. We suggest that this recent colonization is induced by changes in the oxygen levels, which has affected the ecological dominance in this benthic community the same way the bay in Sao Paulo state. Following this finding, we now can prove once again that this species has arrived in Brazil, and it has spread to other estuarine and coastal zones influencing the native fauna.

Presence of specific bio indicators such as *Bulliminela elegantissima* and *Bolivina striatula*, with low oxygen concentrations and high concentrations of organic matter caused by anthropogenic influences in various parts of Florianopolis Bay is a favorable environment for the occurrence of the invasive species *T. hadai*.

# Chapter 7
# Sea Level Forecast Indicators: Sea-Level Rise Forecast by "The Moon and Sun" Foraminifera Species Indicators

**Abstract** Global Sea-Level Rise is caused by two main factors: thermal expansion (ocean water warms and expands), and the contribution from ice sheets (glaciers, land-based ice sheets, and sea ice) due to increased melting, and the continued increase in the twenty-first century is important in determining the distribution of species and habitats, and projections show that by 2100, thermal expansion and glacial melting are expected to cause a rise by 0.26 to 0.98 m, based on climate models, and considering high and low emissions scenarios. However, the contribution of the Greenland and West Antarctic ice sheets could increase these amounts. A potential analogue to modern greenhouse conditions is the climate of approximately five to three million years ago, in the early Pliocene epoch, which provides important clues where earth was in one of the warmest times showing the increase of hurricane pathways all over the world.

The increase in the sea level rate in the Anthropocene Epoch is to blame for higher temperatures. As temperature rises, water heats up and expands causing warming oceans to take up more space and invade land. Higher temperatures melt glaciers and ice caps faster, which in turn adds more water into the oceans. The higher water temperature is likely to increase extreme sea level events, like storm-driven waves, surges, tropical and extra tropical cyclones (also known as hurricanes and typhoons), which have become nearly three times more intense and frequent during the twentieth century. More frequent and stronger hurricanes imply greater heating of water parcels, with consequent warmer temperatures in the tropics, and in turn, more hurricanes.

The total rise of sea level from 1901 to 2010 is larger than the average rate during the previous 2000 years of 19 cm, with global average sea level trends showing an increase by 0.18 cm per year, and the Intergovernmental Panel on Climate Change forecasts that global sea levels will rise by around 47 to 62 cm by 2080.

Foraminiferal assemblages indicate the differences in ocean levels from past to present with the establishment of species zonation occupying marshes, bays, lagoons, reefs, and deep-water marine environments. The rather narrow and accurate sensitivity to limiting factors make Forams "the markers of the sun and the moon" in time and space, and paleo deposits related to paleo-sea level in all ecosystems around the world. Likewise, they forecast future changes and impacts, allowing us to give recommendations to mitigate and remediate effects.

 
P. P. Beck Eichler, C. P. Barker, *Benthic Foraminiferal Ecology*,
https://doi.org/10.1007/978-3-030-61463-8_7

Photo Symbiont-bearing large benthic Foraminifera (LBF) species with zooxanthellae algae which dwells in reef areas needs the sun's proximity under the water to flourish, therefore, light is a limiting factor for such organisms, because these symbionts need to be able to photosynthesize.

The sun and moon phases balance of forces are the largest influencer of marine habitats and their organisms' adaptations imposing nature's cyclical changes by tidal influence, and the most sensitive organisms are capable of registering these rather small-scale yet long-term alterations. The Sun, which is the "star of the show" at the center of our Solar System provides light and heat keeping everything on Earth alive, and it leads to the four seasons of Spring, Summer, Autumn, and Winter, which influence changes in temperature and amount of light on Earth. Environmental indicator organisms sense these changes.

The water above LBF has to permit enough light penetration through the water column for the success of their symbionts, and for reef symbionts, and also optimum abiotic parameters for every specific species in intertidal sites under variations of tidal oscillations. Other Forams from mangroves, estuaries and bays are controlled by a series of oscillating parameters (salinity, temperature, dissolved $O_2$, organic carbon) which often have no direct relationship with actual water depth, yet also serve as precise indicators of sea levels.

Zonation of calcareous and agglutinated foraminifera species depend on the amount of environment freshwater controlled by tide influence. Bay or lagoon environments with marine influence have calcareous forms; agglutinated forms represent brackish lagoons, marshes, and mangroves with less marine influence and higher freshwater influence, and LBF with symbiotic algae in the oceanic realm of reef areas.

Periods of high-frequency sea-level oscillations on the shoreline are confirmed by bottom calcareous bio facies followed by agglutinated ones, being precisely seasonally evidenced. The high-resolution response of microfauna to water level reconstructions based on freshwater and saline balance offers the potential to obtain more detailed and reliable records of relative sea-level change from salt marsh and mangrove sediments.

Sedimentation sites where emergence and submergence have taken place are combined with data to obtain a relatively complete sea-level history by understanding the dynamic of the community of Forams established which permits forecasting future impacts, with only a few as direct "Sea level Forecast" indicators, and have an interesting response in the present, with their ability to foresee future changes, allowing us to recommend actions to prevent losses.

Sea-level rise will directly affect mangroves and beaches where, for example, diverse fauna, which provides turtle nesting, will also be highly impacted by an encroaching ocean. Turtle and seabird nesting beaches are particularly vulnerable to rising sea levels, which will exacerbate beach erosion and flood nests. It is only possible to recommend strategies to protect natural as well as private and public resources with an acute understanding of the environment. With encroaching oceans, mangroves can advance towards land if they are backed by low-lying salt flats as sea levels rise. This can be an adaptation strategy to conserve the vegetation even though the tendency is to lose these important biomes with the existing anthropogenic pressure of its use from public and private infrastructure.

This chapter describes study cases in the south of Brazil in a mangrove area in São Paulo state, then in the reef areas of northeast Brazil, and we finish the chapter with samples collected from Deep Ocean in the north of Australia by the International Ocean Discovery Program (IODP), during Expedition 363, in the West Pacific Warm Pool (WPWP).

With limitations in terms of light, salinity and temperature, the response accuracy of bio indicators is fantastic, and one thing we can be sure of is that with sea-level rising in inundating and eroding coastal habitats, mangrove and reefs may be able to adapt, but only if sea-level rise occurs slowly, and if sufficient sediment exists to continue to accrete vertically to compensate for sea-level rise. These should be of paramount interest in paleoclimate studies, detecting oceanic to freshwater patterns, and reconstructing and forecasting sea levels from fossils and marine cores.

Benthic Forams species for in-situ organic matter flux are a water mass tracer in the northwestern tropical Pacific Ocean, indicating that distribution of lower bathyal and upper abyssal species are separated into two distinct groups. One that reflects relatively warm (>3.5 °C) intermediate depth water, with higher carbon flux (>3.5 g C $m^{-2}$ $year^{-1}$) versus a group that reflects colder, deeper waters with lower carbon flux. This data shows significant correlation with carbon flux and modern bottom water temperatures, reflecting lateral input of organic matter such as associated with down-slope movement. We have evidence for the use of the 'warm' water index as a tracer for water masses, at least in regions where the vertical flux of organic carbon is low and remains low throughout time.

**Keywords** Sea level changes · Symbiont · Mangrove · Salt marshes · Estuaries · Reefs · Deep Ocean · Taphonomy

## Introduction

The continued increase of Sea-Level Rise effects in the twenty-first century is important in determining the distribution of aquatic species and habitats because tides play an important role in foraging and breeding activities. Global Sea-Level Rise is caused by two main factors: thermal expansion (ocean water warms and expands), and the contribution from ice sheets (e.g., from glaciers, land-based ice sheets, and sea ice) due to increased melting. By 2100, projections of Sea-Level Rise shows thermal expansion and glacial melting are expected to cause the sea-level to rise by 0.26 to 0.98 m, based on climate models and considering high and low emissions scenarios, however, the contribution of the Greenland and West Antarctic ice sheet could increase that amount.

Within these scenarios, a potential analogue to modern greenhouse conditions, is the climate of the early Pliocene epoch (approximately five to three million years ago) which can provide important clues to this response in an article written by Fedorov et al. (2010), where earth was warmer, one of the warmest times showing the increase in hurricane pathways all over the world.

The increase in the sea level rate in the Anthropocene Epoch (commencement of significant anthropogenic climate change, but not limited to it) can be blamed by higher temperatures. As temperature rises, water heats up and expands causing the warmer oceans to take up more space and invade land. Higher temperatures melt glaciers and ice caps faster, which in turn adds more water into the ocean. In addition, the higher temperature of waters is likely to increase extreme sea level events like storm-driven waves, surges, tropical cyclones (also known as hurricanes and typhoons), which became nearly three times more intense and frequent during the twentieth century.

These extreme events are part of the Earth's climate system, because by vigorously mixing the upper ocean, they affect the ocean's heat uptake, poleward heat transport, and hence global temperatures, which shows positive feedback between hurricanes and the upper-ocean circulation in the tropical oceans essential to the warm conditions during the early Pliocene Epoch. This feedback is based on the hurricanes ability to warm water that travels toward the Equator at shallow depths and then resurface in other parts of the ocean's wind-driven circulation. These very warm periods contrast to the Anthropocene Epoch of today, but we can see its climate signals of changes appearing. In today's climate, with only a few hurricane tracks intersecting other pathways, there is still little heat exchange between waters. However, if more frequent and stronger hurricanes imply greater heating of water parcels, we will have warmer temperatures in the tropics, and in turn more hurricanes.

The total rise of 19 cm from 1901 to 2010 is larger than the average rate during the previous 2000 years. From 1961 to 2003, the global average sea level trends show increase by 0.18 cm per year. In Australia's Great Barrier Reef, the fastest rates of sea level rise are in the north. The Intergovernmental Panel on Climate Change projects that global sea levels will rise by around 26 to 29 cm by 2030, and by around 47 to 62 cm by 2080.

Foraminiferal assemblages indicate the differences in ocean levels of past and present with the establishment of, and species zonation occupying marshes, bays, lagoons, reefs, and deep water marine environments. They are also capable of forecasting future changes and impacts, allowing us to give recommendations to mitigate and remediate effects.

Foraminiferal assemblage sediments are subject to differentiated bottom water salinities and temperatures due to tidal periodic oscillation derived from moon cycles, which form environmental gradients worldwide within narrowly (around 10 cm in total vertical range) defined vertical zones of occurrence. Reefs and deep water show other abiotic constraints, besides light and tidal influence, such as the amount of necessary oxygen and carbon available for maintaining their life.

Photo Symbiont-bearing large benthic Foraminifera (LBF) species with zooxanthellae algae, which dwells in reef areas needs the sun's proximity under the water to flourish, therefore, light is a limiting factor for such organisms, because the symbiont needs to be able to do their photosynthesis. LBF harbors symbiotic algae called zooxanthellae (dinoflagellates) which are crucial to their hosts survival.

Like leaves in the tree, these microscopic algae harvest light and produce energy in the form of carbon rich compounds for their symbionts. The zooxanthellae, or symbiotic algae, lives within the foraminifera shell. These symbiotic algae remove and use the foraminifera's waste products ($CO_2$, nitrogen) for their growth and photosynthesis. In return, the foraminifera gain energy rich organic carbon, created by the algae through photosynthesis. This symbiotic relationship provides most of the foraminifera's energy needs and contributes to the production of the foraminiferal calcareous skeleton.

The sun and the moon balance of forces are the biggest influencer of marine habitats and organisms adaptations inflicting cyclical changes on nature, imposed by tide movements, and the most sensitive organisms are capable of registering these rather small-scale yet long-term alterations. The Moon is a natural satellite of the Earth, orbiting around it every 28 days. Both Earth and Moon orbit around the Sun, which is the "star of the show" at the center of our Solar System providing light and heat keeping all things on Earth alive. It is the Earth's sidereal movement around the Sun once a year that leads to the four seasons of Spring, Summer, Autumn and Winter, which influences changes in temperature and amount of light on the Earth. Environmental indicator organisms sense these changes.

The sensitivity to limiting factors in marine life are narrow and accurate making Forams "the markers of the sun and the moon" in time and space. These assemblage zones, once discovered in marine sediments, are paleo deposits, and are accurately related to paleo-sea level in all ecosystems around the world.

The key to success of foraminiferal species as sea-level indicators is that their primary controlling factor appears to be elevation above mean sea level, as numerical analyses of faunal assemblages indicate that distribution of taxa is vertically constrained, divided into faunal zones that are mostly elevation dependent.

The water above LBF also has to permit enough light penetrating through the water column for the success of their symbionts, and for reef symbionts, and optimum abiotic parameters for every specific species in intertidal sites, under variations of tidal oscillations. Other Forams from mangroves, estuaries and bays are controlled by a series of oscillating parameters (salinity, temperature, dissolved $O_2$, organic carbon) which often have no direct relationship with actual water depth yet also serve as precise indicators, because under appropriate circumstances information concerning sea level can be derived from other Forams assemblages using indirect means. Sites, which have suffered substantial emergence often contain small basins, which prior to emergence were entirely marine.

The marine sediments at the bottom can be identified using Forams and the sill elevation of the basin can then be related to prior sea-level stands as well. Sedimentation sites where emergence and submergence have taken place can be combined to obtain a relatively complete sea-level history by understanding the dynamic of the community of Forams established. These paleo data patterns provide a framework, which permits forecasting future impacts.

However, among them all, in marine and coastal ecosystems, only a few within a thousand living Forams species are direct "Sea level Forecasters" indicators! This chapter deals with them, and their fantastic responses in the present, with their ability to forecast future changes in a concise detailed yet generalized view!

## *Why Is Important to Measure Sea Level Rise?*

In the northern banks of Buxton in North Carolina, USA, the Cape Hatteras lighthouse, the tallest lighthouse in the US had to be moved 1 km inland in 1999, from its original location, to keep it safe from the encroaching ocean (Fig. 7.1).

The lighthouse was constructed in 1870, and a century later was already being threatened by the ocean. Measures of this type is an example of action to prevent losses.

**Fig. 7.1** (L) The original location of the Cape Hatteras lighthouse. (R) The lighthouse "on the move" in the outer banks of North Carolina. (Photos from old post cards)

## *Gradient of Light, Salinity and Temperature by "Moon and Sun" Indicators*

"Moon and sun" indicators of sea level rise in the present (Holocene) reveals typical assemblage from intertidal and supratidal marsh zones recording different suitable environments of deposition. If there is, for example, a shift of coastline occurring towards land in specific areas in a transgression phase, or the other way around if there was a regression.

Zonation of calcareous and agglutinated foraminifera species depend on the amount of freshwater in the environment controlled by tide influence. Bay or lagoon environments with marine influence have calcareous forms; agglutinated forms represent brackish lagoons, marshes, and mangroves with less marine influence and higher freshwater influence, and larger foraminifera with symbiotic algae (LBF) in the oceanic realm of reef areas. Periods of high-frequency sea-level oscillations on the shoreline can be confirmed by bottom calcareous bio facies followed by agglutinated ones, and can be precisely seasonally evidenced.

We reconstruct changes in water depth from microfossil preserved within mangrove and salt marsh sediments. In the high marsh zone, usually agglutinated foraminiferal species dominate, and in the lower marsh environments, calcareous species are prevalent. Postmortem test dissolution and other taphonomic process that happen after the creature dies can also influence fossil assemblages in many ways, as transport, deposition, exportation, dissolution and burial, among many others, keep occurring and affecting the shell. Sometimes, unfortunately these isolated or combined factors leave no modern analogues.

But when they do leave traces behind, the high-resolution response of microfauna to water level reconstructions based on freshwater and saline balance offers the potential to obtain more detailed and reliable records of relative sea-level change from salt marsh and mangrove sediments.

These sea-level indicators are able to unravel lateral shifts and gradients in depositional environment into a vertical and temporal context. Faunal assemblages, faunal zone boundaries, and peak abundances of agglutinated salt-marsh foraminifera are therefor very precise sea-level indicators.

A monospecific assemblage is one of the most useful data in Holocene sea-level studies. It can be of *Trochammina macrescens* in a ~20 cm vertical range near the upper limit of tidal influence of modern marshes, and peak abundances of *Tiphotrocha comprimata*, found today between mean high water and mean higher high water. These faunal zone boundaries and peak abundances of *Tiphotrocha comprimata* indicate tide levels with an estimated precision of 15 cm or better in the Coast of Maine, USA by Gehrels (1994). According to this author, sea level rose −1.0 mm/year between 5500 and 2500 cal year BP, ~0.5 mm/year in the past 2500 years, and perhaps accelerated in the past 1000 years. Long-term rates of relative sea-level rise for the eastern Maine coast are similar to those from 'stable' coasts and to those previously determined along the central and southern Maine coast. What is happening in Maine is similar to that of adjacent maritime Canada.

Similar assemblages were also observed in Oregon, British Columbia, and throughout New England. The assemblages found in Brazil are applicable to the study of sea-level variation as recorded in Quaternary deposits, like their counterparts in temperate marshes. These studies are well constrained vertical transects of many mangrove faunas in South America.

Data from Australia's fringing mangrove environments done by Woodroffe et al. (2005) support the vertical zonation concept as well, which suggests that the distribution of foraminifera in the intertidal zone is usually a direct function of elevation, with the duration and frequency of subaerial exposure as the most important factor. An agglutinated foraminiferal assemblage dominated by *Miliammina fusca, Trochammina inflata, Ammotium* spp. and *Haplophragmoides* spp. exists at the landward edge of the field sites, in a zone between just above Mean Low Water of Neap Tides to Highest Astronomical Tide level (a vertical range of 1.8 m). In addition, a foraminiferal assemblage dominated by *Ammonia* spp. exists between below Mean Low Water of Neap Tides and Mean High Water of Neap Tides (a vertical range of 0.8 m). These assemblages are used to reconstruct sea level from fossil cores from the Australian area.

This chapter deals with how and why the differences between foraminiferal associations from mangroves, reef areas and the deep ocean are taking place worldwide establishing a baseline of understanding paleoclimate alterations and future sea level changes from these recent very diverse environments.

Our first case study is in the south of Brazil in a mangrove area in São Paulo state. The second case is the reef areas of the northeast of Brazil, and we will finish the chapter with samples collected from Deep Ocean in the north of Australia by the International Ocean Discovery Program (IODP), during Expedition 363, in the West Pacific Warm Pool (WPWP). Each ecosystem has different limitations in terms of light, salinity and temperature; and the response accuracy of bio indicators is fantastic!

## *Mangrove Foraminifera from Bertioga Channel (SP)*

In the mangroves of southwest of Brazil, features such as altitude, rate of sediment transport, and balance between fresh and marine waters by tidal movement provides natural estuarine variability, and where salinity and temperature among other environmental parameters of interface water-sediment influence faunal zones can generally be distinguished.

In Bertioga Channel (São Paulo, Brazil) for example "Higher, Middle, Lower and Marine" foraminiferal zonation of species from two intertidal mangrove ecosystems are distinguished and occur in response to salinity gradients (Fig. 7.2). Samples in the mangrove A were collected 20 m from the estuarine channel. Both samplings were done in sediment of the mangrove's vegetation.

The zonation mainly relates to fresh and marine waters balance, where *Pseudononion atlanticum, Miliolinella sp., Cassidulina subglobosa,* and *Discorbis*

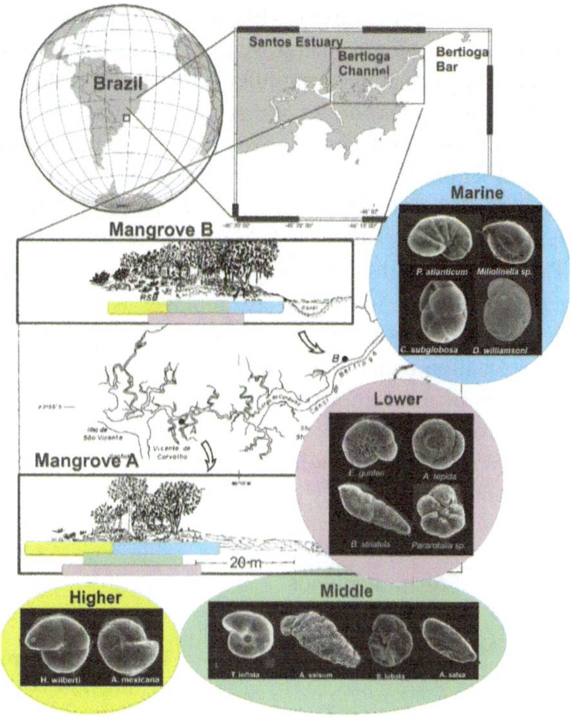


**Fig. 7.2** Marine, Lower, Middle, and Higher Environments from mangroves A and B settings

*williamsoni* inhabit the most "Marine" environments with higher salinities closer to the estuarine channel, whereas "Lower" mangrove environments are dominated by *Ammonia tepida, Elphidium gunteri, Bolivina striatula,* and *Pararotalia sp.* showing preference for marine sites where low freshwater influences the mangroves. The "Middle and Higher" mangrove sites are the preference for all these agglutinated species: *Trochammina inflata, Ammotium salsum, Siphotrochammina lobate, Ammoastuta salsa, Haphlophragmoides wilberti* and *Arenoparrella mexicana.* Among them all, "Higher" mangrove sites with occurrence of *Haphlophragmoides wilberti* and *Arenoparrella mexicana* are the least saline environments where freshwater is very high, and the reach of tides is minimum. Figure 7.2 shows differentiation between the mangrove located closer to the northern mouth of Bertioga Channel as the most saline environment with higher numbers of dead calcareous species, evidencing greater tidal influence than the mangrove located near the southern mouth. Mangrove A presents more restricted environments with microfauna characteristic of a more brackish freshwater influence than mangrove B. This detailed observation of faunal composition and the association with specific mangroves trees roots shows that Mangrove B differs from A specifically because of the occurrence of dead Miliolina (Forams with thicker shells), and some species are not mutually common.

Samples collected in the roots of mangrove tree *Laguncularia racemosa* show no Miliolina Forams species, whereas in the roots of *Cipriacea* and of *Rhizophora mangle* (mangrove tree) Milioline totaled only 2%. Muddy substrates from *Avicennia germinans* mangrove trees are restricted environments which harbor low diverse agglutinated foraminiferal assemblage, dominated by *Haplophragmoides wilberti* and *Arenoparella mexicana* in the most elevated landward area. The assemblage composed of *Throchamina inflata* and other agglutinated taxa only is indicative of High Mean Water Neap Tide elevations. A moderately diverse mixed assemblage of agglutinated and calcareous foraminifera inhabits sites around *Laguncularia racemosa* and represents the Mean Low Water Neap Tide elevation. The assemblage dominated by species of *Arenoparella mexicana*. The species richness gradient of individual assemblages is tide-related revealing strong trend of decreasing diversity moving up the shore. These distributions are similar to intertidal zonation patterns in mangroves and estuaries and highlight their potential for sea level reconstructions. Sea level elevations are well recorded by the upward transition from *Rhizophora* - dominated intertidal mangrove mud to freshwater swamp (clearly identified by pollen analysis) and by the basal contacts of beach barrier sediments which sharply overlie the upper shoreface.

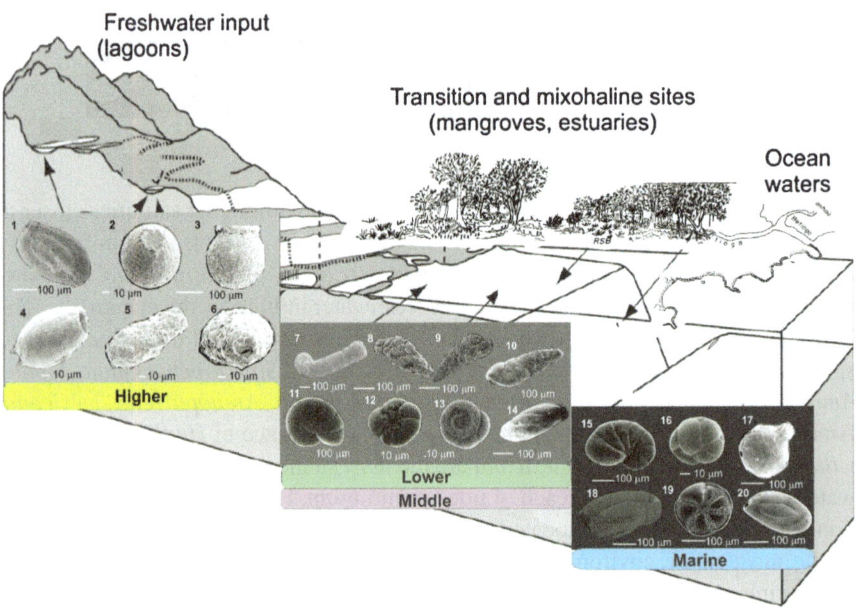


**Fig. 7.3** Differentiation on the microfossils species zonation. Higher freshwater input where the highest influence of tide reach inside the lagoons with occurrence of *Miliammina fusca* (1) and thecamoebians, the true freshwater forms (2–6). Transition and mixohaline sites of mangroves and estuaries in the Lower and Middle parts have as dominant species (*Ammobaculites exiguus,* 7; *Ammotium salsum,* 8; *Gaudryina exillis,* 9; *Bolivina striatula,* 10; *Elphidium poyeanum,* 11; *Ammonia tepida* (12, ventral view, 13 dorsal view); and *Buliminella elegantissima* (14). The ocean waters with marine sites have as representative fauna species of *Pseudononion atlanticum,* 15; *Cassidulina subglobosa,* 16; *Diatrona sp,* 17; *Quinqueloculina lamarckiana,* 18; *Buccella peruviana,* 19; and *Quinqueloculina patagonica,* 20

This kind of species zonation can be extrapolated to geographically larger environments, and it is possible to describe very accurately the zonation of different sub environments within its species community specific characteristics. Figure 7.3 shows schematic environments from lagoons to ocean waters and the foraminiferal fauna associated.

The robust schema on the gradient of microfossils species zonation where the highest freshwater input is also the highest influence of tide which are able to reach the insides of lagoons. Species characteristics of this sub environment is *Miliammina fusca* (1) with the true freshwater forms, the thecamoebians. Sub environments of mangroves and estuaries, which are transition and mixohaline sites in the Lower and Middle parts have *Ammobaculites exiguus*; *Ammotium salsum, Gaudryina exillis, Bolivina striatula, Elphidium poyeanum, Ammonia tepida*, and *Buliminella elegantissima* as dominant species.

The influence of marine waters from the adjacent ocean shows that *Pseudononion atlanticum,* 15; *Cassidulina subglobosa,* 16; *Diatrona sp,* 17; *Quinqueloculina lamarckiana,* 18; *Buccella peruviana,*19; and *Quinqueloculina patagonica,* 20 are the representative fauna of these sub environments.

This zonation of sub environments are common worldwide with slight species changes, and provides a scheme that is applicable to short-term sea-level change studies in many sites worldwide and are easily compared because of the general comparable environmental natural settings.

These conditions factor the estuarine circulation, making the north part more conductive to the invasion of the tide, and in the southern areas of the channel, less accessible to the entrance of the tides due to its attenuation, affects the mangroves in different ways. Largo do Candinho has null velocities, as the tidal waves and currents that enter in opposite directions through the two mouths meet there, forming the "fall of the waters" encounter of the tides, as demonstrated by Harari and Camargo (1998).

This fact causes the water to move at speeds lower than the critical erosion velocities, which minimizes the effect of the sedimentary transport of the bottom, characterizing the Bertioga Channel as a corridor of suspended sediments. It is observed, therefore, that the estuarine circulation is conditioning the deposition of sediments and shells of foraminifera. The influence of the tide along the Bertioga Channel is also observed in the estuarine Channels and adjacent mangroves, through dominant foraminiferal species, being indicative of zonation of microenvironments.

The differentiation of the two mangroves showed zonation of agglutinated species, subject to tidal influence, evaporation and Channel geometry. *Haplophragmoides wilberti* proved to be characteristic species of brackish environments (less saline) than *Arenoparrella mexicana*.

Sea-level rise will directly affect mangroves, and beaches where diverse fauna provide turtle nesting will also be highly impacted by the encroaching ocean. Turtle and seabird nesting beaches are particularly vulnerable to rising sea levels, as this would exacerbate beach erosion and flood nests.

From understanding the environment, and how it works, it is possible to recommend strategies to protect nature itself, as well private and public resources. It is therefore imperative to protect mangroves that are backed by low-lying salt flats, because in the case of an encroaching ocean, the mangroves can advance towards land, as sea level rises. This can be an adaptation strategy to conserve the vegetation even though the tendency is to lose these important biomes with the already anthropogenic pressure of its use in public and private infrastructure.

With sea-level rise, inundating and eroding coastal habitats, they "may" be able to adapt, only if sea-level rise occurs slowly with enough time for them to adapt, and if sufficient sediment exists to continue to accrete vertically to compensate for sea-level rise. Focusing on deltas and other areas of accretion so that mangroves can colonize and expand is mandatory. Prioritizing turtle beaches backed by coastal plains, rather than narrow pocket beaches which are likely to erode or get covered by waves and high tides, might also improve the likelihood of persistence of sea turtles and other species as well over time. The urban adequate expansion of spaces needs to occur respecting environmental laws of protection and conservation of species.

The physical and geomorphological characteristics of the estuarine syste affect estuarine circulation patterns and, consequently, zonation of Foraminiferal species between marine, mixohaline, brackish and continental faunas are gradational and very specific. In rivers, estuarine channels and mangroves, where the environments are considered mixohaline, species of agglutinated foraminifera as *Arenoparrella mexicana, Haplophragmoides wilberti, Ammotium salsum, Gaudryina exillis* and *Trochammina inflata* were observed. Hyaline genera as *Ammonia* and *Elphidium* and Milioline as *Quinqueloculina* were rare because acidic conditions prevent the construction of their shells. These forms are mostly dead indicating that these shells were transported.

We conclude that the response of the fauna is so accurate that it should be of paramount interest in paleoclimate studies, detecting oceanic-to-freshwater patterns and reconstructing and forecasting sea level from fossil and marine cores.

## *Impacts on Reef Ecosystems Tracked by Larger Benthic Forams (LBF) and Their Light's Limitations*

As discussed above, the global average sea level has risen over the last half century by about 2–3 mm per year. Even with many reefs that are currently subjected to tidal regimes of several meters, based on sea-level rise rate, the impacts on coral reefs will only occur if the projected rate and magnitude of sea-level rise are within limits of potential growth rate, which is the accretion of sedimentary rates. Even though the rapid growth rate of many corals (e.g., 2 to >30 mm/year) is enough to correlate well with sea-level rise projections, the existing threats to reefs from increasing sea temperatures, ocean acidification, disease, and overfishing reduce the ability of corals to maintain healthy organisms. In particular, ocean acidification may significantly slow both rates of coral growth and reef accretion, making it even more difficult for corals to grow and flourish.

The increase of sedimentary processes rises sea levels locally, because of interference with photosynthesis, feeding, recruitment, reproduction, propagules emission, and other key physiological reef processes. The increased sedimentation due to shoreline erosion can smother reefs or reduce sunlight needed for photosynthesis and symbionts. For example, the large and fast 6 m sea-level rise during the last interglacial led to smothering and rapid retreat of reef structures all over the world.

Even small increases in sea level of 0.2 m are able to increase turbidity on fringing reefs which in turn creates an increase in re-suspension of fine sediment on reef flats, or an increase in coastal erosion and transport of fine sediment to adjacent reefs. Sea levels had been very constant for the past 6000 years, therefore, sea level rise is predicted to increase at a higher rate, and coral reef growth may not be able to keep pace. The shape and existence of some coastlines, bays and islands may also be affected.

Because much of the land adjacent to Reefs and other inundated ecosystems is low-lying, small changes in sea level will mean greater erosion and land inundation, with less light arriving to the reefs. This will also cause significant changes in tidal habitats such as mangroves, and move saltwater into low-lying freshwater habitats, as already mentioned earlier in this chapter. They are all connected, and this will also have flow-on effects for numerous other juvenile species that use these habitats for protection and food resources.

It is believed that most reefs will probably be able to cope with the current rate of sea level increase because the maximum rate of reef growth is in average twice this rate. With the increasing in sea temperatures, ocean acidification, diseases, and overfishing as additional challenges, coral reefs therefore also face the rise of sea level impacts. Rising greenhouse gas concentrations are causing the average global ocean temperature to rise, however, global warming is also causing water levels to rise due to melting ice and increased thermal expansion of seawater.

According to models forecasting future climate presented by the recent studies done by the Intergovernmental Panel on Climate Change (IPCC), the annual rate of water level rise, currently equivalent to 3 mm per year since 1993, may double or more by 2100 (to more than 6 mm per year). They predict a global rise in sea level from +0.40 m to over +0.82 m by 2100, according to the most pessimistic scenarios. The extent of coastal change and coastal vulnerability to extreme events will depend directly on maintaining reef elevation relative to water levels. Indeed, the reef structure serves as a natural barrier against erosion by reducing wave energy and regulating water circulation—a service that they would no longer be able to provide by the end of the century.

The growth rates of coral reefs in the Atlantic and Indian Oceans show that even under a modest climate change scenario (2.6 °C increase in atmospheric temperature, 3% and 6% of the coral reefs studied in the Indian and Atlantic oceans respectively) would be able to compensate for projected sea-level rise. For the most pessimistic cases, most of the reefs studied would experience a rise in water depth of more than 0.5 m, exacerbating the destructive impact of waves on reef areas.

Ecological and social consequences are also expected because of a change in the hydrodynamic regime at the level of the reef ridge which can threaten the ecosystems of lagoons seagrass beds—important areas for the establishment and development of young organisms (juvenile species). The rise in the average water level will also cause greater coastal erosion, favoring flooding of the land during extreme phenomena, as well as a reduction in land uses and fresh water reserves due to salinization of the soil and groundwater. A socioeconomic loss as well, since more than 275 million people live in and or within 30 km of reefs, so the socio-economic impacts will be catastrophic, forcing a significant number of local communities to abandon their lands, on a scale of a few decades within the near future. However changes in depth-controlled physical processes will be minimal to non-detectible in all aspects of reef morphology of colony, including sizes and shapes, adaptability, cross-reef relief, and how they are able to maintain healthy organisms, meeting their needs, and keeping up with the pace of rising sea levels.

Rates of vertical reef accretion in Hawaii and throughout the Pacific are smaller, and so in the order of magnitude of predicted rates of sea-level rise, unfortunately it is unlikely that reefs there and in other similar locations will keep pace, so their inability to cope will lead to subtle but important changes in their structure and processes. Besides this, another challenge faced by coral reefs in Hawaii is the high flux of colder and enriched in terrestrial-derived nutrients; submarine groundwater discharge from land is very different from surrounding original seawater. This input is likely to buffer thermal stress and supply them with nutrients from oligotrophic coastal waters. Thermal stress imposed by warming waters is responsible for bleaching in corals and LBF with symbiotic forms that use photosynthesis as their way to obtain energy. Together with an excess of terrestrial-derived nutrients this causes loss of coral and LBF, with replacement by macroalgae and other opportunistic species. Changes in warming and acidification of ocean waters, as well as sea water rise, and land uses that alter quantity, quality, flux, composition, and rate of groundwater, especially in rapidly developing areas.

— These are near future problems that need to be addressed now, before it is too late —.

Submarine groundwater discharge can shape coral reef ecosystem structure and the ecosystem services they provide, so detailed studies on their mapping is important for addressing challenges on these ecosystems.

Sea level rise influences coral reefs and they respond to rapid sea-level rise at a decadal time-scale. The increased wave energy and altered circulation across reefs is sure to affect circulation and sediment deposition, nutrient, contaminant, and larval dynamics with propagules emissions from higher diverse ecosystems to less diverse ones.

Our approach requires an understanding of the physical parameters of sediment, and human-induced impact in the foraminiferal community, and the resulting reef ecosystem, so we can better predict what will occur due to climate change. Interdisciplinary studies try to answer questions relying on a combination of field sampling and measurements using a wide range of parameters including oceanographic instruments (acoustic Doppler current profilers, wave/tide gauges, tempera-

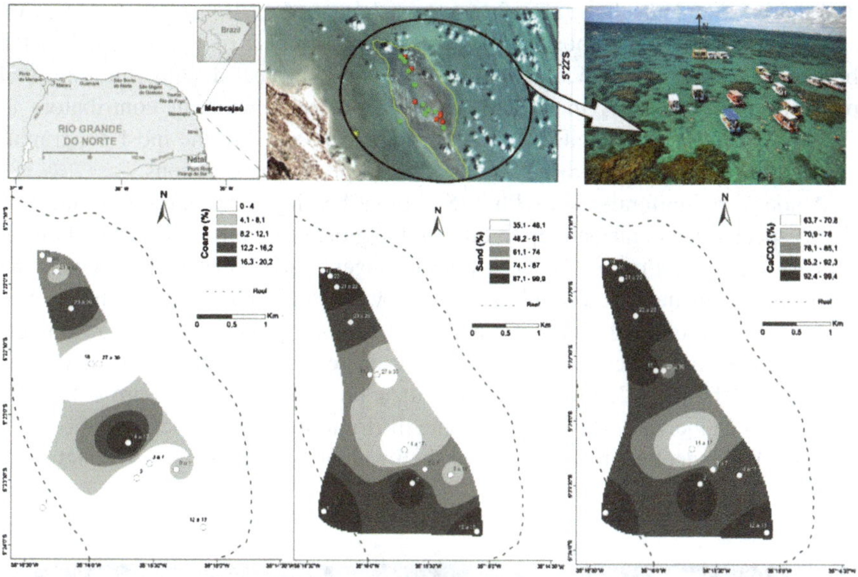


**Fig. 7.4** Maracajau reefs and maps of coarse sediment grains, $CaCO_3$, and sand concentrations in the sediment

ture sensors, salinity sensors, chemical sensors) on the seabed or on moorings, water-column profilers with similar suites of sensors, coral and sediment cores, geophysical water-column, and sub-bottom surveys. Maps, of coarse sediment grains, $CaCO_3$, and sand concentrations in the sediment are illustrated on Fig. 7.4 and they are useful to provide insight on how processes on reefs may respond to predicted future sea-level rise.

These distributions of coarse, sand, and $CaCO_3$ in the sediment from Fig. 7.4 show that central and northern parts of Maracajau reefs have higher values of coarse sediments showing higher hydrodynamics preventing deposition of fine sedimentary facies; also it shows exporting and rather eroding parts of the ecosystem in these mentioned sites.

Sand and $CaCO_3$ are accumulating in the south and north parts, but in the central parts sand and $CaCO_3$ declines and coarser grains dominate. There is a buildup of sand and coarsest sediments all over the reefs showing high currents and velocities allowing heavier grains to settle closer to their origin source, and preventing fine sediments from depositing in these parts. The distribution of $CaCO_3$ show that in the central parts where coarse grain size sediment predominates, there is less bio contribution for the $CaCO_3$ factory. Maps for data visualization are useful to provide insight on how processes on reefs may respond to predicted future sea-level rise.

The decline in calcite saturation reduces calcification and increased dissolution rates of major marine calcifiers such as corals, algae, foraminifera, and some mollusks, with potentially severe implications for coral reef ecosystems. A lack of sufficient explanation about calcification mechanisms explaining species-specific differences stimulates the study of the establishment of relationship between $pCO_2$ and calcification. It is important for assessing the impact of lowering pH in the future, and the possible ecological consequences for reef-dweller foraminiferal populations when $pCO_2$ increases, since Forams are important contributors to $CaCO_3$ (calcite and aragonite) cycling in the ocean, and, are the most geographically widespread group of calcifying organisms, producing skeletons.

Among foraminiferal species, Photo Symbiont-bearing large benthic Foraminifera (LBF) species can contribute up to 80% of the global foraminiferal reef carbonate, approximately 1 kg/m$^2$ of $CaCO_3$ a year. Changes in pH affect LBF survival, and calcification causing severe symbiont loss known as 'bleaching', already mentioned earlier. The presence of symbiotic algae, an ecological and evolution advantage, since increase in $CO_2$ concentration in seawater releases algae from the dissolved inorganic carbon (DIC) limitations, and their photosynthetic rates are enhanced, resulting in an increase of pH levels within the foraminifer's microenvironment. On the other hand, decline in pH decreases shell weight in general, causes reduction in

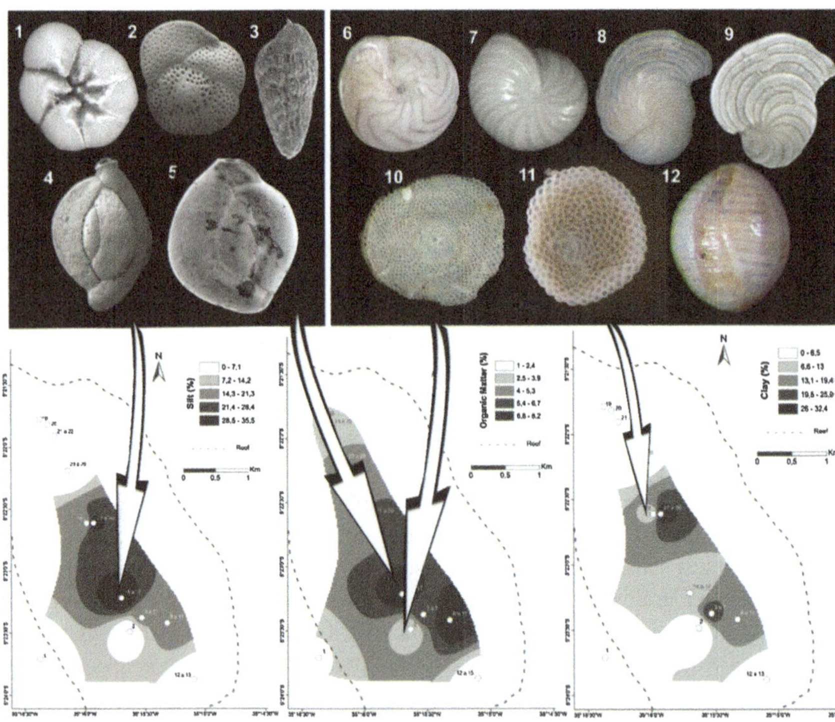


**Fig. 7.5** The distribution of fine sediments in this figure illustrates distributions of silt, clay and organic matter, and the small opportunistic species with higher organic matter and silt (1. *Ammonia tepida,* 2. *Discorbis* peruvianus, 3. *Bolivina striatula,* 4. *Quinqueloculina patagonica,* 5. *Quinqueloculina lamarckiana*) and the LBF (6. Amphistegina gibbosa, 7. *Peneroplis carinatus* 8. and 9. *Laevipeneroplis proteus* 10. and 11. *Amphisorus hemprichii,* 12. *Borelis melo*)

calcification rates, and photosynthetic efficiency, and alter shells through dissolution of calcium carbonate crystals decreasing the $CaCO_3$ factory.

The distribution of fine sediments in Fig. 7.5 illustrates distributions of silt, clay, and organic matter. Deposition of them follows similar patterns of deposition of fine sediments in the central parts showing buildup of fine material that are prone to accumulate and serve as food for marine species. The low velocity of currents within inner parts of reef corals shows protected areas that should serve as fine sediments sinking in higher depths.

The small individuals of Forams species are both marine and opportunistic, preferring environments of muddy sediment and high organic matter (1. *Ammonia tepida*, 2. *Discorbis peruvianus*, 3. *Bolivina striatula*, 4. *Quinqueloculina patagonica*, 5. *Quinqueloculina lamarckiana*) and the LBF 6. *Amphistegina gibbosa*, 7. *Peneroplis carinatus* 8., and 9. *Laevipeneroplis proteus* 10. and 11. *Amphisorus hemprichii*, 12. *Borelis melo*) dominating where there are larger grain sizes and clay.

Different settings of this reef result in two contrasting styles of sedimentary sequences, one exposed moderate energy with larger clayed sediment grains and larger Forams, and a protected, low energy, muddy with organic matter and silt in the inner parts of the reefs with the opportunistic fauna. The erosion and deposition patterns are exposed and it can be represented by Forams species. The silty clay rich in organic detritus in inner deposited silts and sands; is cyclic and is controlled by turbidity currents and water mass flows. Near-bottom process as redistribution and deposition by contour-following bottom currents, are responsible for deposition of the sediment.

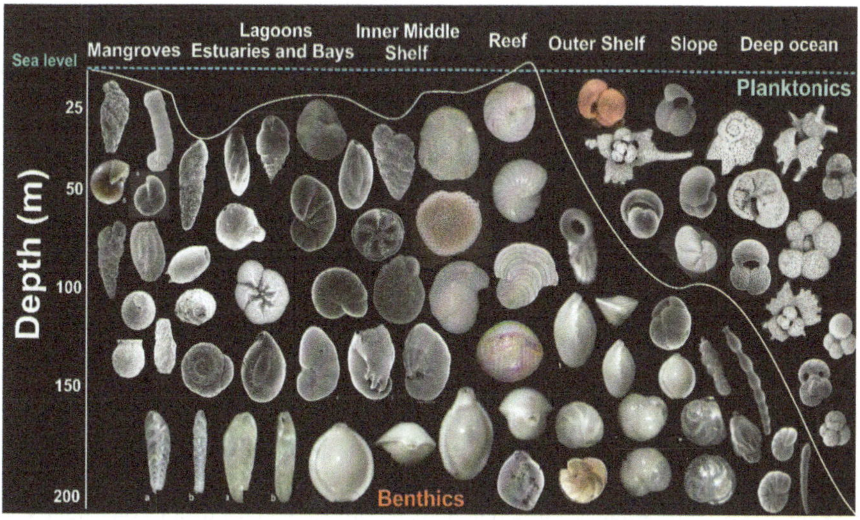


**Fig. 7.6** A scheme with dominant benthic and planktonic species in mangroves, lagoons, estuaries, bays, and in the shelves (inner, middle and outer), reefs, slopes and the Deep Ocean. A sea level line shows environments below sea level, with exception of mangroves, and reefs areas, which would be at risk for sea level rise

Since recent water masses and geo properties are recorded in Forams shells and each environment has specific conditions that influences the community dynamics, a very general yet detailed schema on the marine environments and its Forams species can be detailed in Fig. 7.6.

Using these microfossils it is possible to show the evolution of environments and ecosystems in the present, having an insight on near future and problems in mangroves, lagoons, estuaries, bays, or any kind of shelf (inner, middle and outer), reefs, slopes and the Deep ocean, depending what species they are harboring. An observation on the sea level line shows that all environments are located below sea level, with exception of mangroves, and reefs areas, which would be then more susceptible and exposed to sea level rises.

A scheme like that is useful because it can reflect glacial-interglacial climatic fluctuations of the past. In the previous glacial phases, the continental shelf was emergent, and river sediments were discharged directly into submarine canyons in reef from where they could easily be transported downslope to the continental margin and abyssal plains. In contrast, during the relatively short intervals (5000 to 20,000 years ago) of high sea level that accompanied the Holocene (125,000 to 115,000 years B.P.) in the beginning of the Last Inter Glaciation and previous interglacial phases, the locus of river sedimentation migrated landward across the continental shelf. The low gradient and great width of the shelf, plus strong longshore currents prevented seaward movement of sediment beyond the innermost shelf. Using Forams communities dynamics we can unravel this kind of valuable information.

## *Warm (Shallow) and Cold (Deep) Species from Deep Ocean in the North of Australia*

Periods of shallowing environments within deepening trends in the marine realm is a general temporal succession feature of benthic foraminiferal assemblages. Differentiate assemblages occur when intensification of the oxygen minimum zone inhibited a diverse benthic community for example, and a monospecific fauna dominates in a too warm, too enriched with high organic flux ecosystem, or when fauna shows preference for a very oxygenated and colder environments.

Abundant high organic flux revealed by faunal zones of bio indicators can indicate past environments or past events, for example, an active upwelling of cold enriched nutrients water mass during any specific time and space.

Forams as Bio indicator of temperature can differentiate between warmer, intermediate water assemblages ('warm' water group) versus a colder, upper deep-water assemblage ('cold' water group). The two groups of species differ consistently in carbon flux and bottom water temperatures.

With that in mind, the last example of this chapter deals with marine sediment cores from the Northwestern tropical Pacific Ocean, in the Western Pacific Warm

Pool (WPWP), sampled with R/V "JOIDES Resolution" by the International Ocean Discovery Program (IODP) Expedition 363 which reveals cool or warm phases of faunal fluctuations related to the present and past. These findings are an opportunity to investigate climate changes of the last 20 million years.

Correlation with water depth reflects the foraminifera's specific habitat. This is primarily defined by food availability (organic matter, or carbon, flux) and dissolved oxygen, which are associated with primary productivity at the surface and its respiration in the water column, at the sea bed, and in pore waters (e.g., Schmiedl et al. 1997; Sen Gupta 1999; Katz et al. 2003; Jorissen et al. 2007). The depth at which specific water masses dominate depends on the salinity and temperature (i.e. density), and therefore benthic foraminiferal assemblages should also relate, although circumstantially, to these parameters.

The fluctuations we will show and tell are characterized by species arrangement in the same faunal groups used by Lutze (1977, 1979) and Lutze and Coulbourn (1984) with very good correlation between a warm indicative group of benthic species and the climatic curve derived from planktonic foraminifers by Pflaumann (1975).

The fluctuations uses the total proportion of "warm" benthic species (*Laticarinina pauperata, Cibicidoides kullenbergi, C. robertsonianus, Cibicidoides sp., Hoeglundina elegans,* and *Bulimina aculeata*) expressed in percentages of both "cool" species groups (*Pyrgo murrhina, Cibicidoides wuellerstorfi, Uvigerina peregrina,* and *Globobulimina hoeglundi*) whereas percentages of all other species are ignored. The resulting curve is called "the Warm Benthic Foraminifers (WBF) curve".

These warm benthic Forams species assemblages from core tops were used by Billups et al. (2020) as tracers for carbon flux and modern bottom water temperatures in intermediate to deep-water masses, in the WPWP in the present time, suggesting that this warm and cool species index is accurate to trace temperature (Fig. 7.7). The high significance of the correlations at our study sites supports that the 'warm' water index may provide a tracer for temporal changes in carbon flux.

The authors mentioned above have separated the assemblage into two groups. One of the groups reflecting relatively warm (>3.5 °C) intermediate depth water with higher carbon flux (>3.5 g C $m^{-2}$ $year^{-1}$) where *Bolivina robusta, Bulimina aculeata, Globobulimina pacifica, Hoeglundina elegans, Laticarina pauperata,* and *Cibicidoides pachyderma* thrives versus a group that reflects the colder, deeper waters with lower carbon flux where *Oridorsalis umbonatus, Uvigerina bifurcata,* and *Planulina wuellerstorfi* dominate.

It is interesting to notice that down-core variations of *Planulina wuellerstorfi,* shows this species as more prevalent in areas of lower organic carbon flux rates (Altenbach et al. 2003). The amount of organic matter reaching the sea bed depends on respiration in the water column, and the amount reaching the sea bed, decreases with water depth (e.g., Berger and Wefer 1991; Suess 1980) establishing a first order link between benthic foraminiferal species and bathymetry. On the other hand, *Uvigerina proboscidea,* a species associated with high carbon flux, correlates well with high organic carbon in the past, and the warm species, supporting the use of this species and the warm and cool species index as a productivity proxy, even when carbon flux is relatively low and *U. proboscidea* is absent.

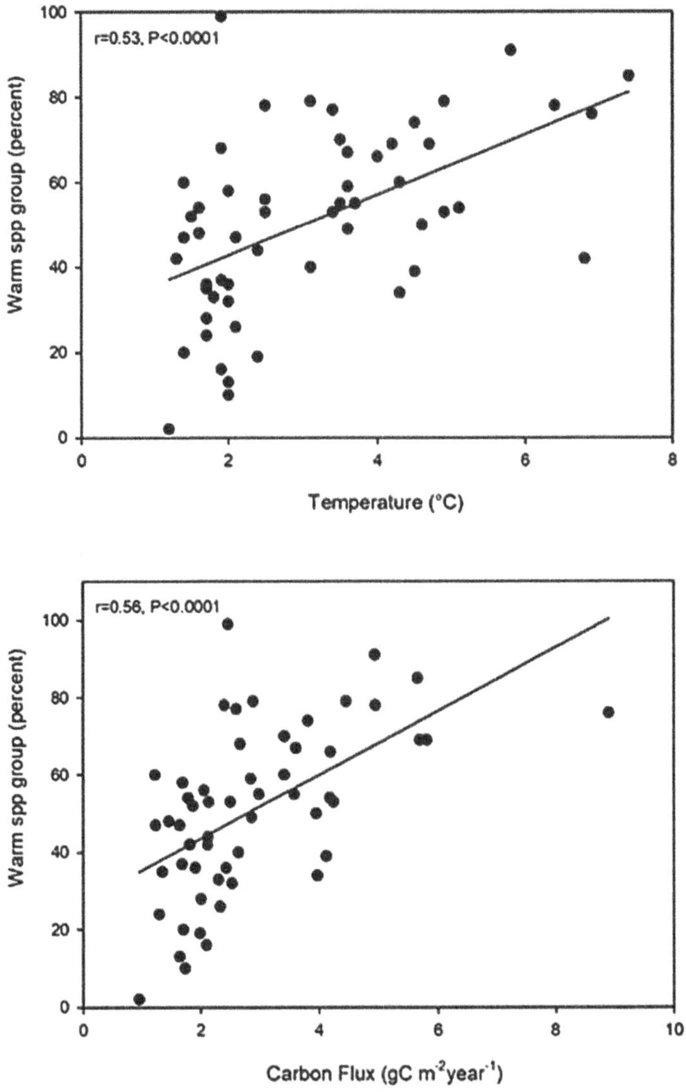


**Fig. 7.7** Carbon flux and modern bottom water temperatures in intermediate to deep-water masses, in the WPWP in the present time, versus warm species group (Billups et al. 2020)

The use of benthic foraminiferal species as a water mass tracer in regions wherein primary productivity and carbon flux remain low shows the reach of bio indicators even when food is not available. The ability to show warm (shallow) and cool (deep) water balance in microfossils preserved from more than 30 million years ago into the Miocene provides an interesting framework of all past fluctuations of temperature and food availability, and link it to sea level rise and the amount of ice in the world at a specific time in the past. An interesting example of an occurrence of species from reef areas in the Papua New Guinea in areas of 1000 m depth remains unanswered and it will be explain in the last topic of this chapter.

## *The Paleo Submerged Reef in the Papua New Guinea*

Samples from Papua New Guinea, a tectonically complex region close to the shelf break (Rosenthal et al. 2017) contain abundant specimen of reef (*Amphisorus hemprichii, Coscinospira arietina, Operculina complanata, and Peneroplis planatus*) and shallow (>200 m water depth, *Rotalinoides compressiusculus, Bulimina marginata* and *Elphidium advenum*) benthic foraminiferal species (Rosenthal et al. 2017). *Bulimina marginata:* It is rarely encountered in deeper bathyal assemblages; and it is usually from lagoonal areas, these fauna are found at water depths less than 45 m (Van Morkhoven et al. 1986; Jones and Brady 1994). *Hyalina balthica*: It is a neritic upper bathyal species (~400–1000 m). *Bulimina aculeata*: It is an ecophenotype of *Bulimina marginata* from upper depth limits within the upper bathyal zone to the abyssal zone (deeper water).

Hole U1484A differs from Holes U1482A and U1483A due to the more common occurrence of benthic foraminifers, particularly between Samples 363-U1484A-7H-CC and 17H-CC (62.27–156.48 mbsf) where planktonic to benthic foraminifer ratios reach as much as 70:30. Changes in benthic foraminifer assemblages are noted in the core catcher samples, with the upper and lower successions dominated by deep-water benthic foraminifer fauna, and the middle section consisting of mixed deep and shallow water benthic foraminifer fauna, including reef-dwelling taxa. The assemblages are discussed in detail below.

Benthic foraminifera fauna from Samples 363-U1484A-1H-CC to 6H-CC (5.00–52.95 mbsf) and 17H-CC to 27F-CC (156.48–223.62 mbsf) are composed mostly of the following deep-water benthic foraminifer fauna: *Cibicidoides pachiderma, Uvigerina hispida, Uvigerina prosbocidae, Bolivinita quadrilatera* and *Bolivina robusta*. These species are identified as bathyal forms, occurring at depths of up to 3500 m (Van Morkhoven et al. 1986; Jones and Brady 1994). In Samples 363-U1484A-7H-CC to 9H-CC (62.27–81.21 mbsf) and 15H-CC to 17H-CC (138.62–156.48), benthic foraminifer fauna is composed of a mixture of the deep water fauna mentioned previously and the shallow water taxon *Rotalinoides compressiusculus*, which are usually found at 4–70 m water depths (Jones and Brady 1994).

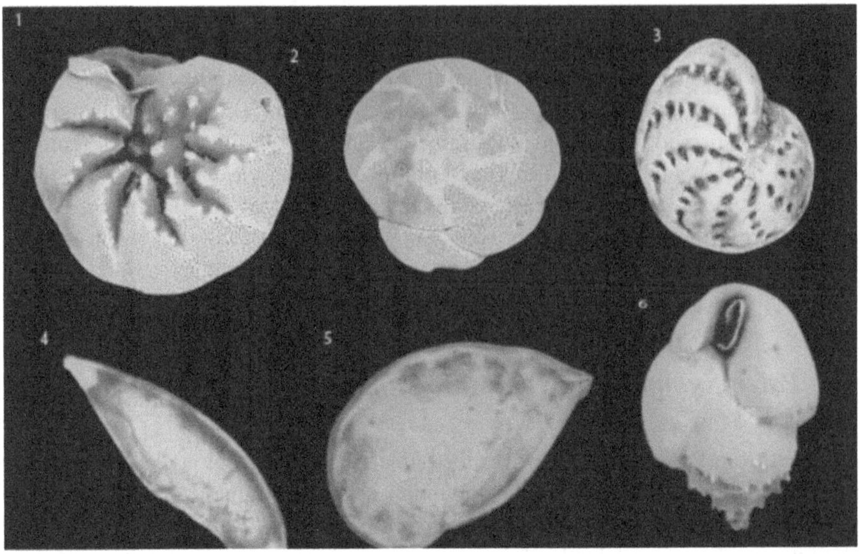

**Fig. 7.8** Benthic Foraminifera species show a fauna transition Shallow water species (1, 2, 3, 6)—Reef-dwellers species (4, 5, 7, 8, 9, 10)—Deep water species (11, 12, 13) that is indicative of sea level rise. 1. *Rotalinoides compressiusculus* (Brady 1884) (ventral view) 363-U1484A-15HCC. 2. *Rotalinoides compressiusculus* (dorsal view) 363-U1484A-15HCC. 3. *Elphidium advenum* (Cushman 1922) 363-U1484A-5HCC. 4. *Planularia australis* (Chapman 1915) 363-U1484A-8HCC. 5. *Planularia australis* (Chapman 1915) 363-U1484A-8HCC. 6. *Bulimina marginata* (D'Orbigny 1826) 363-U1484A-11HCC. 7. *Peneroplis planatus* (Fichtel and Moll 1798) 363-U1484A-15HCC. 8. *Coscinospira arietina* (Batsch 1791) 363-U1484A-15HCC. 9. *Amphisorus hemprichii* (Ehrenberg 1839) 363-U1484A-15HCC. 10. Operculina complanata (Defrance 1822) 363-U1484A-15HCC. 11. *Bolivinita quadrilatera* (Schwager 1866) 363-U1484A-4HCC. 12. *Bolivina robusta* (Brady 1884) 363-U1484A-4HCC. 13. *Bulimina aculeata* (D'Orbigny 1826) 363-U1484A-3HCC

Reef-dwelling benthic foraminifer fauna were observed between Samples 363-U1484A-10H-CC and15H-CC (90.58–138.62 mbsf). The fauna consists of *Operculina complanata* and rare specimens of *Amphisorus hemprichii, Peneroplis planatus* and *Coscinospira arietina.*

Benthic Foraminifera species show a fauna transition shallow water species (1, 2, 3, 6)—Reef-dwellers species (4, 5, 7, 8, 9, 10)—Deep water species (11, 12, 13) that is indicative of sea level rise as seen in Fig. 7.8.

# Chapter 8
# Volcanic Past Cycles Indicators: Paleoclimatology and Extinctions Using Benthic and Planktonic Forams Community Dynamics

**Abstract** The Benthic and planktonic foraminiferal communities' dynamics as volcanic past cycles indicators are very well placed within the Paleoclimatology and extinctions studies. We have showed a bit, of what is available to explain how communities have evolved in the past. The past volcanic activity has released as much carbon dioxide into the atmosphere as anthropogenic as predicted emissions projections for the twenty-first century and they are linked to increases in carbon dioxide emissions and with faunal patterns, with marine extinctions observed sediment cores after volcanic episodes, and this increase in carbon dioxide and other volcanic gaseous influences on global warming and ocean acidification is responsible for the extinction of three quarters of species on Earth on the past. For example, dinosaurs were pretty much extinct because of "The Deccan Traps", an igneous province, one of the largest volcanic features on Earth, located on the west-central India, and the Siberian Traps have influenced the end-Permian extinction, in which more than 90% of life on Earth disappeared. Many patterns should be first understood to be able to forecast future climate change scenarios. We can however explain that the modern ongoing carbon dioxide emissions are similar to those that led to the end-Triassic mass extinction. The importance of understanding Earth's deep water past is predicated on predicting how it will respond to future climate change. The mass extinction and high-stress conditions were explained by the intense Deccan volcanism leading to rapid global warming and cooling, with enhanced weathering, continental runoff, and ocean acidification, resulting in a carbonate crisis in the marine environment. The chronic explosive volcanic activity generated unstable benthic habitat colonized by only a few species. The increase in atmospheric $CO_2$ concentrations lead to decreased pH and carbonate availability in the ocean, known as Ocean Acidification, and the ability of marine invertebrates to tolerate acidity are the 'windows into the future' to study. Cores with ashes and tephra in Papua New Guinea (PNG) during Expedition 363 sampled by the IODP show that total foraminiferal diversity was low when volcanic activity was in place detected by the presence of tephra and volcanic ashes. Foraminiferal density and diversity in PNG were high and similar to those observed on the Great Barrier Reef or other sites, however diversity decreases, and show inverse correlation by benthic foraminifera to high presence of ashes and tephra in the past. However, ecological studies from shallow

 
P. P. Beck Eichler, C. P. Barker, *Benthic Foraminiferal Ecology*,
https://doi.org/10.1007/978-3-030-61463-8_8

reef environments observed increased foraminiferal dominance of opportunists when corals became rare from chronic or acute anthropogenic influences, for example with sewage and oil spills. Agglutinate taxa that do not rely on calcification will replace calcifying species, and we call it a fauna replacement by invasive species. Density and diversity of agglutinated taxa is also in decline, but are less marked than calcifying taxa in an environment where pH is low. Dissolution of foraminifera seen in marine sediment under elevated $pCO_2$ unravels other direct ecological impacts. Impacts such as dissolution and loss of biogenesis of carbonate by other organisms that are under near-future $pCO_2$ conditions, which will reach a problematic real-time scenario. None of the previous extinctions were as severe as the ecological or even taxonomic extinction in shallow carbonate areas which we are predicting. Because of the rate of increasing $pCO_2$, and unfortunately, we expect that the increase in the temperature in the Holocene and the tendency until 2100 will take us to the warmest Pliocene climate with the unfortunate consequences of living in a warmer than optimum world. The variability based on the frequency and intensity of some events are one of the warmest our world has ever seen, reflecting changes in temperature derived from data from deep sea sediment core samples, and of course shells of benthic and planktonic Forams and other organisms like pollen act as proxies in drilled marine sediment cores reflecting historic climate. A unique fauna of foraminiferal species from these highly opposed environments created by differences in temperature in the past are recorded paleo cycles, of which responds to the amount of ice in the world, due to their high sensitivity to the environmental changes in the modern and past sediments. Here we show that tephra and ashes of IODP Hole U1485A (Exp. 363 WPWP) record a periodicity of explosive volcanism within the last 0.8 Myr. Possible triggering mechanisms for these mass flow deposits include earthquakes and associated tsunamis and shelf/slope sediment instabilities during times of rapid deposition such as can occur during river flood events. Over longer timescales, it is also possible that sea level played a role in the storage and release of sediment from the PNG shelf (although the shelf itself is very narrow) and from the paleo-valley of the Sepik River, which is a relatively large area presently few meters above sea level. Changes in diversity shows balance of alternating deep (cold) and shallow (warm) benthic foraminifera fauna along time in the past. The “at least” five decreases in diversity peaks in the past show that the response of the benthic community to adverse climate is a change in their ecological pattern. These changes can take a whole community and an entire ecosystem to extinctions, and we have already seen five extinctions along Earth’s history. And if history teaches us anything, it is how to react to and prepare for crisis rather than repeat mistakes. Research suggests we are fast approaching disastrous effects of this sixth Anthropocene extinction. However, we can successfully surmount the challenges of biodiversity loss and climate change and dramatically alter the trajectory if we can pinpoint and remediate problems within a near future. With our planet “in crisis”, evidence demonstrates widespread ecological collapse and biodiversity loss. We know that as average temperatures rise and the frequency of extremely warm years increases, the impacts of habitat loss and fragmentation become even more increasingly apparent. We are with without a doubt entering a sixth mass extinction event because of the rapid decline in biodiversity. The majority of these species inhabit environmentally delicate tropical and subtropical areas susceptible to human impacts. This refers to a situation where the extinction of one species affects other species that rely on it for survival, thereby also placing them at a 'domino effect' risk of extinction as part of a destructive chain reaction. Stop cutting and burn forests, stop global trade of wild species, study and protect, preserve, and conserve our planet's biodiversity.

**Keywords** Volcanic eruptions · Carbon dioxide · Extinctions · Organic material · Temperature · Climate change

## Introduction

Volcanic activity pulses release huge quantities of greenhouse gases from Earth's inner system. Past volcanic activity has released as much carbon dioxide into the atmosphere as the anthropogenic predicted emissions projections for the twenty-first century. A single volcanic pulse of activity during the end of the Triassic (202 million years ago), for example released as much carbon dioxide as humans are expected to emit over the course of the twenty-first century. Volcanic Eruptions are linked to increases in carbon dioxide emissions, and with faunal patterns, extinctions are observed in marine sediment cores from volcanic episodes.

The increase in carbon dioxide and other volcanic gaseous influences on global warming and ocean acidification are responsible for the extinction of three quarters of Earth's species. The massive environmental and climatic changes created by eruptions are associated with the timing of mass extinctions and other changes. The formation of large igneous provinces are followed by shifts in climate and faunal patterns, and mass extinctions. The dinosaurs were pretty much extinct because of "The Deccan Traps", a large igneous province located on the west-central India (17–24°N, 73–74°E), one of the largest volcanic features on Earth, whereas the Siberian Traps is thought to have influenced the end-Permian extinction, in which more than 90% of life on Earth disappeared.

Estimating carbon dioxide emissions for ancient eruptions is a challenge. Because Carbon is a volatile element, it readily degasses during magma rise and eruption. Moreover, carbon may be added to old rocks due to alteration. Besides that, there are innumerous variables that should be first understood to be able to forecast future climate change scenarios. We can however explain that the modern ongoing carbon dioxide emissions are similar to those that led to the end-Triassic mass extinction.

The importance of understanding Earth's deep water past is predicated on predicting how it will respond to future climate change. Scientists use models based on decades of historical weather and climate data to predict future climate change. Voluminous data is available to look at Earth's paleo patterns of four and half billion year history of climate change, to provide insight into modern climate change, which includes volcanic past cycles and extinctions of benthic foraminiferal community dynamics as indicators. Here we will show a bit of this available data to explain how communities have evolved in the past.

## *Volcanic Eruptions Responsible for Ocean Acidification Related to Extinction of Forams*

A correlation between Deccan eruptions and mass extinctions reveals that around 50% of plankton foraminifer species gradually disappeared during volcanic eruptions prior to the first of four lava mega flows, reaching out to the Bay of Bengal, from around 1500 km within India. Moreover, the other 50% disappeared after the first mega flow, and the mass extinction was complete with the last mega flow. Throughout these intervals, blooms of opportunist species dominated shallow-marine assemblages that are analogous worldwide, from India to the Tethys, and the Atlantic Ocean to the Pacific. Similar high-stress environments dominated by blooms of opportunistic species correlating with Deccan volcanism phase 3, in the early Danian, was followed by full biotic recovery after volcanism ended. The mass extinction and high-stress conditions may be explained by the intense Deccan volcanism leading to rapid global warming and cooling, with enhanced weathering, continental runoff and ocean acidification, resulting in a carbonate crisis in the marine environment.

Chronic explosive volcanic activity generates an unstable benthic habitat that is colonized by only a few species. In a volcanically active crater at the NW Rota-1 Seamount, microbial mats and populations of mobile vent shrimp and crabs are the only persistent organisms (Embley et al. 2006). Lava flows, mass wasting, and explosive eruptions destroy benthic communities (Haymon et al. 1993). These events create a 'time zero', a Krakatau-like reset of the system (Whittaker et al. 2010), with well-documented 'recovery' of communities to pre-disturbance conditions taking place well within a decade (Shank et al. 1998; Tunnicliffe et al. 1997).

The increase in atmospheric $CO_2$ concentrations lead to decreased pH and carbonate availability in the ocean, known as Ocean Acidification. The ability of marine invertebrates to tolerate acidity are the 'windows into the future' to study.

We have studied benthic foraminifera in sediments from cores with ashes and tephra in Papua New Guinea (PNG) during the Expedition 363 sampled by the IODP, and we observe that total foraminiferal diversity was low when volcanic activity was in place, detected by the presence of tephra and volcanic ashes. Foraminiferal density and diversity in PNG were high and similar to those observed on the Great Barrier Reef or other sites, however diversity decreases and show inverse correlation by benthic foraminifera to high presence of ashes and tephra in the past.

Diversity gradients in the PNG cores suggested that all common Forams species responded to the cyclic pulses of volcanic eruptions in the past, and its unstable ecological conditions created by the increase of the $CO_2$ in the past. If an ocean is still warming, it is likely to occur an increase in $CO_2$, which in turn, except under the most optimistic scenario, will extinct all calcareous species at the $CO_2$ conditions predicted for the year 2100. We fear that the best scenario is unlikely to occur, because it requires large and immediate emission cut-offs and negative net emission (carbon sequestration) at least until the end of this century. In addition, mass extinctions of benthic foraminifera from deep sea have occurred in the past, most of which were related to temperature, and of course the increase in $pCO_2$.

Ecological studies from shallow reef environments observed show increased foraminiferal dominance of opportunists when corals became rare from chronic or acute anthropogenic influences such as sewage and oil spills for example. However, an important microhabitat and food source for a few foraminiferal species opportunists is where phytodetritus aggregates from domestic and industrial effluents. Phytodetritus has its origin in the euphotic zone and quickly settles to the seafloor after the spring bloom. Small individuals of opportunists are adapted to a fluctuating nutrient supply, and able to grow and reproduce when these suitable phytodetritus food sources are present, and the distribution and abundance of these opportunists is controlled by presence of organic material on the seafloor, from these food like phytodetritus deposits. Fortunately, nowadays fluctuations in the abundance of some specific foraminiferal phytodetritus related species are concomitant where millennial-scale and high-latitude climate perturbations data is available to study.

Agglutinate taxa that do not rely on calcification will replace calcifying species, and we call it a fauna replacement, which was already discussed, in an earlier chapter about invasive species. Density and diversity of agglutinated taxa also decline but are less marked than calcifying taxa in an environment where pH is low. However, uncommonly agglutinated taxa in marine waters leave open niches by past calcifying taxa. We still do not know what the impact these can cause to the whole fauna community.

Dissolution of foraminifera seen in marine sediment under elevated $pCO_2$ reveals other direct ecological impacts. Impacts such as dissolution and loss of biogenesis of carbonate by other organisms that are under near-future $pCO_2$ conditions, which will reach a problematic real-time scenario. None of the previous extinctions were as severe as the ecological or even taxonomic extinction in shallow carbonate areas, which we are predicting because of the rate of increasing $pCO_2$.

Previous natural $pCO_2$ increases occurred in one to two orders of magnitude slower, and were associated with low pH and less reduced calcite or aragonite saturation states than the anthropogenic increases presently discussed. Our data suggests that the induced ecological extinction of shallow tropical benthic foraminifera by year 2100, will be similar to the five other mass extinctions observed in the geological past.

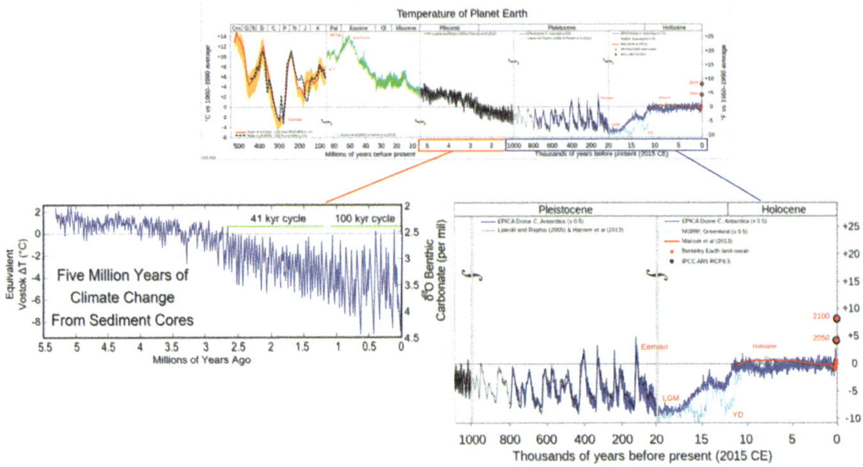


**Fig. 8.1** Evolution of temperature of Planet Earth until 2100. A detail of the last 5.5 million years and from 1000 years before present (compiled and modified from Alley (2000), Zachos et al. (2001) and Lisiecki and Raymo (2005))

In Fig. 8.1 we show how the temperature of Planet Earth has changed along its 500 million years until present based on Alley (2000), Zachos et al. (2001) and Lisiecki and Raymo (2005). It is possible to observe variability during 5.5 million years in the Pliocene, where a globally synchronous onset of glacial marine isotope oxygen stages (MIS) was described linking the frequency and intensity of some events to atmospheric perturbations, such as inter-hemispheric climate variability in one of the warmest older periods. More recently in the Pleistocene, a millennial-scale change in atmospheric $CO_2$ also had an impact on the overall pattern of abundance of specific species tolerating low pH and high detritus in a dark world under intense volcanic activity. In the end of the Pleistocene, the lower temperature marked by the Last Glacial Maximum also had a significant effect on the species distribution. We expect that the increase in the temperature in the Holocene and this tendency until 2100 will take us to the warmest Pliocene climate, with the unfortunate consequence of living in a warmer than optimum world.

## *Glacial and Interglacial Cycles Within Marine Isotope Oxygen Stages (MIS)*

Marine isotope stages (MIS), marine oxygen-isotope stages, or yet oxygen isotope stages (OIS), in a globally synchronous onset of glacial marine isotope oxygen stages within that variability, reflects alternating warm and cool periods in the Earth's paleoclimate during 5.5 million years in the Pliocene. The variability based on the frequency and intensity of some events are one of the warmest our world has ever seen, reflecting changes in temperature derived from data from deep sea sediment core samples, and of course from shells of benthic and planktonic Forams!

Working backwards from the present (which is MIS 1 in the scale), stages with even numbers have high levels of oxygen-18 and represent cold glacial periods, while the odd-numbered stages are troughs in the oxygen-18 figures, representing warm interglacial intervals in 800Ka within the Pliocene. The data derived from pollen and foraminifera (plankton) remains as proxies in drilled marine sediment cores (sapropels) reflecting historic climate (Fig. 8.2).

These highly opposed environments created by differences in temperature in the past recording paleo cycles, give conditions to the development of a unique fauna of foraminiferal species which responds to the amount of ice in the world, due to their high sensitivity to the environmental changes in the modern and past sediments.

The Western Pacific Warm Pool (WPWP), one of the warmest marine waters of the world nowadays, is an interesting place to evaluate special limits and characteristics of the paleo record to highlight the influences that the temperature differences are acting on in the onset of MIS. This onset of the MIS Stages of marine core sections from the South Atlantic are about a million years old. The following are the start dates (apart from MIS 5 sub-stages) of the most recent MIS (Lisiecki and Raymo 2005; LR04 Benthic Stack). The figures are in thousands of years. Numbers for substages in MIS 5 denote peaks of sub stages rather than boundaries.

The MIS timescale was developed from the pioneering work of Cesare Emiliani (1954a, b, 1956, 1966) and is nowadays widely used in archaeology and other fields to express dating in the Quaternary period (the last 2.6 million years). They provide the fullest and best paleoclimatology data for that period range, with the opportunity to study the early climate of the Earth, representing "the standard to which we correlate other Quaternary climate records". Emiliani's work in turn depended on Harold Urey's prediction in a paper from 1947 that the ratio between oxygen-18 and oxygen-16 isotopes in calcite, the main chemical component of the shells and other

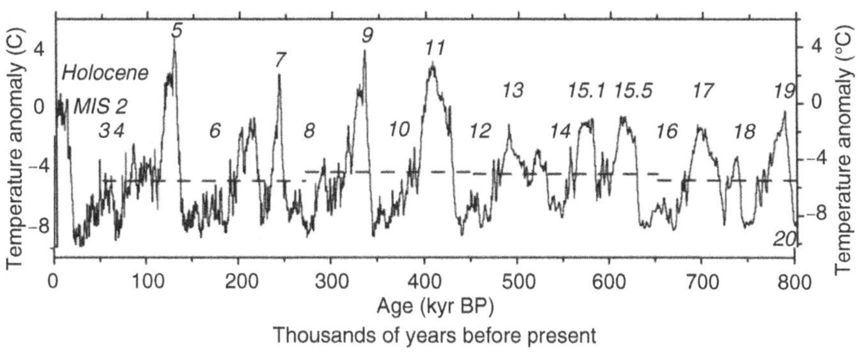


**Fig. 8.2** Globally synchronous onset of glacial marine isotope oxygen stages (MIS) reflects alternating warm and cool periods in the Earth's paleoclimate, during 800 Ka in the Pliocene. MIS 1 is the present in the scale, stages with even numbers have high levels of oxygen-18 (cold glacial periods), while the odd-numbered stages are troughs in the oxygen-18 figures (warm interglacial intervals)

hard parts of a wide range of marine organisms, should vary depending on the prevailing water temperature in which the calcite was formed.

Over 100 stages have been identified, currently going back some six million years, and the scale may reach back up to 15 Mya. Some stages, in particular MIS 5, are divided into sub-stages, such as "MIS 5a", with 5 a, c, and e being warm and b and d being cold. A numeric system for referring to "horizons"(events rather than periods) may also be used, for example with MIS 5.5 representing the peak point of MIS 5e, and 5.51, 5.52 etc. representing the peaks and troughs of the record at a more detailed level. For more recent periods, increasingly precise resolution of timing continues to be developed.

The MIS data also matches the astronomical data of Milankovitch cycles of orbital forcing or the effects of variations in insolation caused by cyclical slight changes in the tilt of the earth's axis of rotation—the "orbital theory". The marine sediments also acquire depositional remnant magnetization which allows them to be correlated with earth's geomagnetic reversals. For older core samples, individual annual depositions cannot usually be distinguished, and dating is taken from the geomagnetic information in the cores. Other information, especially as to the ratios of gases such as carbon dioxide in the atmosphere, is provided by analysis of ice cores.

## *MIS Paleo Thermometers ($\delta^{18}O$ Mg/Ca and Sr/Ca)*

The isotopic ratio of $^{18}O$ to $^{16}O$ from Forams shells or ice cores shows mean temperatures in the past. High values mean low temperatures. More ice means higher $\delta^{18}O$ values. Ocean temperatures are relatively stable and the $\delta^{18}O$ value reflects the temperature where precipitation occurs.

Magnesium (Mg) is incorporated into the calcite shells (tests) of planktonic and benthic foraminifera as a trace element. Because the incorporation of Mg as an impurity in calcite is endothermic, more is incorporated into the growing crystal at higher temperatures. Therefore, a high Mg/Ca ratio implies a high temperature, although ecological factors may confound the signal. Mg has a long residence time in the ocean, and so it is possible to largely ignore the effect of changes in seawater Mg/Ca on the signal. Strontium (Sr) incorporates in coral aragonite, and it is well established that the precise Sr/Ca ratio in the coral skeleton shows an inverse correlation with the seawater temperature during its bio mineralization.

## *Pleistocene Eruptions*

In the 2.588 ± 0.005 million years BP, approximately when the Quaternary period and Pleistocene epoch begin, massive volcanic eruptions of River flood basalts released $CO_2$ into the atmosphere and triggered a decline in ocean ph. With global temperatures rising because of this, sea levels also rose, flooding large areas of the continents. This created ideal conditions to bury large amounts of carbon from the accumulations of marine organisms into sediments, and to transfer volcanic carbon from the atmosphere to the ocean over tens of thousands of years. Past large episodes of volcanism throughout Earth's history have been linked to mass extinctions and widespread oxygen depletion in the oceans.

The elevated marine productivity and carbon burial helps remove some of the carbon dioxide from the volcanoes and has acted as a negative feedback, mitigating some, but not all, of the climatic effects associated with the outpouring of volcanic $CO_2$. This is why we have an opportunistic fauna dominating when volcanic pulses occur.

Causes for extinctions vary across taxa and by region. In some cases, extinctions appear to be linked to hunting, while in others they are consistent with the ecological effects of climate change, and in a final case a combined effect of hunting and climate change are thought to be responsible. The main good effect that volcanoes have on the environment is to provide nutrients to the surrounding soil. Volcanic ash often contains minerals that are beneficial to plants, and if it is very fine ash, it is able to break down quickly into the soil.

The existence of causal links between volcanism and Earth's climate remains highly controversial, partly because most related studies only cover one glacial cycle. However, longer records with high resolution and time precision are available from marine sediment profiles in which the distribution of tephra records frequency changes of explosive arc volcanism is evident. Here we show that tephra of IODP Hole U1485A (Exp. 363 WPWP) record a periodicity of explosive volcanism within the last 0.8 Mya. Possible triggering mechanisms for these mass flow deposits include earthquakes and associated tsunamis and shelf/slope sediment instabilities during times of rapid deposition such as can occur during river flood events, supporting the idea that silt and sand fractions at Site U1485 are being transported from the Papua New Guinea (PNG) shelf.

Over longer timescales, it is also possible that sea level played a role in the storage and release of sediment from the Papua New Guinea shelf (although the shelf itself is very narrow) and from the paleo-valley of the Sepik River, which is a relatively large area presently a few meters above sea level. This idea is supported by Aiello et al. (2019) which have showed that climate, sea level, and tectonics are controlling factors on the sediment discharge from the Sepik River (PNG) during the Mid-to Late Pleistocene.

In addition to rapid mass flow deposition of coarse sediment to Site U1485, the ubiquitous presence of varying amounts of silt-sized siliciclastic particles within the hemipelagic clay also suggests that other transport pathways contribute to the flux of terrigenous material to Site U1485.

The winnowing of silt-sized particles from the shelf by seasonal and inter annual changes in the strength and direction of the New Guinea Coastal Current (NGCC) and the New Guinea Coastal Undercurrent (NGCUC) is one possible source of fine sediment to Site U1485 (Ueki et al. 2003).

Seasonal variations in river discharge and in the spatial extent of the sediment plume in combination with NGCC and NGCUC dynamics may provide another source of fine sediments.

The change in sedimentation patterns could correspond to changes in the abundance of terrigenous source material, as sand, arriving to the site, possibly due to uplift of the mountains, within the Sepik-Ramu watersheds occurring from tectonic activity on the Bewani-Torricelli and Ramu-Markham faults (Webster et al. 2004). A change in precipitation intensity in the Sepik-Ramu River watersheds, and/or a change in the currents responsible for dispersing the sediments in the ocean are the consequences. Yet another hypothesis, which does not exclude the previous one, is that the change in sedimentation corresponds to a change in location of the point source of coarse sediments; in other words, in the past, the mouth of the Sepik River might have been located more inland and sediment accumulated within the paleo-Sepik embayment.

The occurrence of 25 tephra layers observed throughout Site U1485 (Fig. 8.3) and the occurrence of ash as a minor component mixed with silt and clay minerals indicate the influence of pyroclastic sedimentation throughout the entire depositional history of the site. The tephra layers consist of vesicular glass fragments, bubble wall shards, and micro-pumice, with glass fragments most abundant in tephras in the upper part of the site. The basal contact of the tephra layers varies from

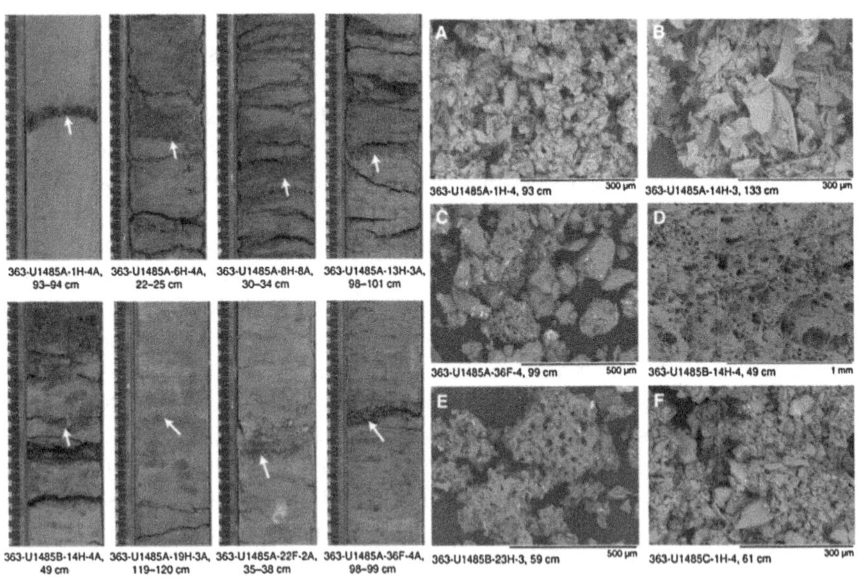


**Fig. 8.3** Tephra layers observed throughout Site U1485, and the occurrence of ash as a minor component mixed with silt and clay minerals indicate the influence of pyroclastic sedimentation throughout the entire depositional history of the site. The thickness of the tephra layers ranges from 0.5 to 4.0 cm, with maximum grain sizes ranging from 60 to 250 µm. Tephra records maintains a strong 100 kyr periodicity (Rosenthal et al. 2017)

sharp to gradual depending on the degree of bioturbation and also possibly current winnowing processes.

At Site U1485, the continuous accumulation of hemipelagic clay and biogenic components is interspersed with more rapidly deposited coarser grained sand, and silty sand. Fluvial clastic shelf sediment is carried to Site U1485 by mass gravity flow (Rosenthal et al. 2017).

Possible triggering mechanisms for these mass flow deposits include earthquakes and associated tsunamis and shelf/slope sediment instabilities during times of rapid deposition such as can occur during river flood events. There is presence of volcanic ash in some intervals of the U14845A core. The 34Hcc, 35Hcc, 36Hcc and 39Hcc intervals present ashes, tephra, and we can pinpoint changes in the fauna pattern at least after these volcanic episodes recorded.

## *Planktonic Foraminifers Fauna and Age at Site 363-U1485A (Rosenthal et al. 2017)*

Sample 1H-1 from mudline at the very top of the core U1485A was examined and contains a typical diverse open-ocean tropical assemblage. Although the marker "in time" species *Globorotalia truncatulinoides* is absent down to Sample 7H-CC (65.01 mbsf), Samples 1Hcc to 4Hcc (7.53–36.56 mbsf) are assigned to the uppermost part of Subzone Pt1b (<0.07 Ma) based on the absence of *Globigerinoides ruber* (pink) and *Globorotalia flexuosa*. Biohorizon top *G. flexuosa* (0.07 Ma) is observed between Samples 4Hcc and 5Hcc (36.56–45.99 mbsf).

Biohorizon top *G. ruber* (pink) (0.12 Ma) is well defined between Samples 8Hcc and 9Hcc (75.02–84.49 mbsf). Biohorizons base *Globigerinella calida* (0.22 Ma) and base *G. flexuosa* (0.40 Ma) are difficult to locate precisely because of the rarity of the markers and because distinguishing both forms from related species requires fine taxonomic discrimination. The former biohorizon is recognized between Samples 15Hcc and 16Hcc (138.99–148.77 mbsf) and the latter between Samples 38Hcc and 39Hcc (273.27–277.78 mbsf). The bottom of the hole at Sample 44Hcc (300.75 mbsf) is found to be still within Subzone Pt1b (<0.61 Ma) based on the absence of *Globorotalia tosaensis*, the marker for the top of Subzone Pt1a. This is confirmed by the fact that *Pulleniatina* populations have a strong dextral dominance throughout the hole, indicating that the entire succession lies above the L1 sinistral excursion (<0.80 Ma).

Figure 8.4 shows Benthic Foraminifera diversity along biozones defined by Planktonic Forams—the "markers of age". The decrease in Benthic diversity along different times and especially after Pleistocene volcanic eruptions is recorded after the presence of volcanic ashes and tephra in some intervals (34Hcc, 35Hcc, 36Hcc and 39Hcc) showing changes in the fauna pattern at least after these volcanic episodes where we spot at least two decreases in benthic diversity after these Pleistocene volcanic eruptions. The benthic Forams diversity curve for site U1485A in Fig. 8.4,

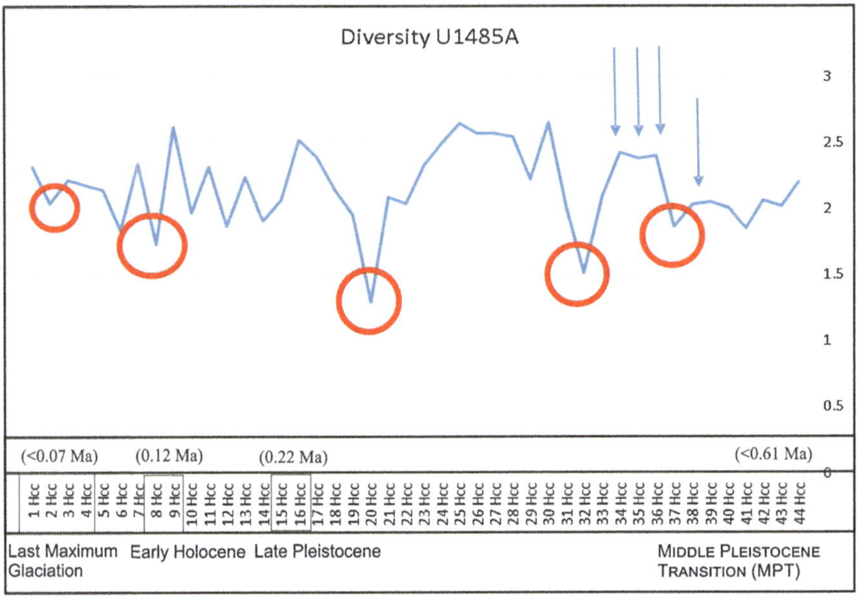


**Fig. 8.4** A benthic Forams diversity curve on the Biozones markers of age for U1485A. Samples 1Hcc to 4Hcc (<0.07 Ma) marks the Last Maximum Glaciation. Samples 8Hcc and 9Hcc (0.12 Ma) record the Early Holocene, Samples 15H-cc and 16Hcc (0.22 Ma) shows the Late Pleistocene, and the 44Fcc (<0.61 Ma) is younger than the Middle Pleistocene Transition (MPT). The volcanic ash in 34Hcc, 35Hcc, 36Hcc, and 39Hcc show at least two decreases in benthic diversity after Pleistocene volcanic eruptions

show samples 1Hcc to 4Hcc (<0.07 Ma) marking the Last Maximum Glaciation, samples 8Hcc and 9Hcc (0.12 Ma) recording the Early Holocene, samples 15H-cc and 16Hcc (0.22 Ma) unravelling the Late Pleistocene, and the 44Fcc (<0.61 Ma) is younger than the Middle Pleistocene Transition (MPT).

## *Benthic Foraminifera Fauna from Site 363-U1485A*

Changes in diversity shows balance of alternating deep (cold) and shallow (warm) benthic foraminifera fauna along time in the past. All deep and shallow water species characteristics are cited below and were described by Van Morkhoven et al. (1986), and Jones and Brady (1994).

The oscillations in their communities dynamics records paleoclimate variation, where deep-water fauna is dominated by *Cibicidoides pachiderma* (Rzehak 1886), an upper bathyal species, from the early Oligocene through the Pleistocene, *Uvigerina hispida* (Schwager 1866) from early Miocene through Pleistocene, and *U. prosbocidae* (Schwager 1866) from late Oligocene through Pleistocene). *Bolivinita quadrilatera* (Schwager 1866) is characteristic of 200–500 m depth is a deep water species and *Bolivina robusta* (Brady 1881) from 3 to 900 m can be found also in shallow water.

In shallow waters, *Uvigerina mediterranea*, an outer neritic upper bathyal species characteristic of warm waters, high salinities, little dissolved oxygen, and high levels of organic matter (Schumacher et al. 2010), together with *Rotalinoides compressiusculus* (Brady 1884), a warm water species, characteristic of very shallow depths (2–37 m) are present in the intervals of 13Hcc and 17Hcc, showing higher diversity peak in interglacial cycle (Fig. 8.4).

The "at least" five decreases in diversity peaks in the past show that the response of the benthic community to adverse climate is a change in their ecological pattern. These changes can take a whole community and an entire ecosystem to extinctions, and we have already seen five extinctions along Earth's history. Are we going to face another one? Is perhaps a sixth Mass extinction coming? In our present time? Will we make geological history?

## *The Anthropocene 6th Extinction: Prediction*

Here we wish to discuss some history up until this day, and segue into our current human environmental effects.

If history teaches us anything, it is how to react to and prepare for crisis rather than being doomed to repeating mistakes. In terms of preparing ourselves, what should we do with what we have learned? Do we want to be polarized in times of crisis or do we want to have unity? Do we want cultural changes imposed upon us, or do we want to make personal and social changes of our own accord?

Things are changing now in terms of what we want personally, and what kind of society and natural world we wish to live in.

Our oceans, lakes and rivers are increasingly empty of life because of humanities greedy little mindedness. Research suggests we are fast approaching disastrous effects of this sixth Anthropocene extinction. However, we can successfully surmount the challenges of biodiversity loss and climate change, and dramatically alter the trajectory for what R. Buckminster Fuller called "Spaceship Earth". His concept is for a comprehensive planetary planning that resulted in new strategies intended to enable all of humanity to live with freedom, comfort and dignity, without negatively affecting the earth's ecosystems or regenerative ability.

# Chapter 9
# Optimum Biodiversity as Climax Indicators: Higher Biodiversity of Large Benthic Symbiotic Foraminifera as Means of Propagules Emission

**Abstract** Numerical analysis and FORAM index (FI) have similar results on the sedimentary reefs health of RN where high hydrodynamics and other reworking processes might influence faunal occurrence of Forams in this portion of the shelf and therefore alter the interpretation of the FI index. However, dominance of smaller taxa, including stress-tolerant species, and minimal representation of symbiont-bearing taxa, indicate unsuitable conditions for foraminifera in Pirangi and in Maracajaú, Rio Grande do Norte (RN), Brazil. Our work also demonstrates that Forams fauna correlates well with coarse and sand fractions, and depth was the least influencing variable and that FI provided reliable proxy for water quality and reef health at RN, but may not be suitable for not typical sedimentary sites. The genus *Amphistegina* occurred mostly in ($10 \leq x \leq 15$ m) and ($15 \leq x \leq 25$ m) domains, and Cluster analysis revealed clear patterns of grouping sites among stations, and could be linked to ecological indices differences. Cluster identified clear separation within the data where sites are grouped in relation to its indication of conduciveness to reef growth or not after event of stress.

Percentage of sand availability and $CaCO_3$, as well as percentage of silt and clay for both reef areas also contributed to the occurrence of other heterotrophic (*Bolivina, Elphidium, Nonion*) and stress-tolerant foraminifera (*Cibicides, Discorbis, Miliolinella, Pyrgo, Quinqueloculina, Rosalina,* and *Triloculina*). In Maracajau, TOM contributed to the occurrence of heterotrophic and stress-tolerant Forams genera, and the low frequency of stained foraminifers in both reef areas, especially in Pirangi, might be related to empty tests common situation, in reef areas, and that most reef-dwelling foraminifers live attached to reefs, stalks and macroalgae, and only empty tests should be found free in the sediment after a reproductive event.

The mean FI values at Pirangi ranged from less than 2 to 4 at most stations, indicating that Pirangi is a marginal environment for reef growth and unsuitable for recovery after stress events according to the index interpretation especially in areas with touristic activities. Coarse and sand fraction, are the controlling parameters. It is also clear that depositional energy plays an important role in transportation and deposition of sediments and Forams. The dominance of *Quinqueloculina* (hetero-

 
P. P. Beck Eichler, C. P. Barker, *Benthic Foraminiferal Ecology*,
https://doi.org/10.1007/978-3-030-61463-8_9

trophic) shows changes that have yet to be identified are occurring in the area and must be taken into consideration in further studies.

According to the FI results, the environmental health evaluation shows that the water quality of the Pirangi is not suitable for coral reef growth and Maracajau has sites suitable for coral reef growth even if it has individual sites where coral could not survive after a stress event.

Long-term assessments are needed in order to improve our knowledge regarding the distribution and ecological importance of Brazilian reef-dwelling foraminifers, as well as to extend the application of the FI to large-scale monitoring of this and other reef ecosystems in the Southwestern Atlantic. Species from RN reefs have taxonomic affinity with all other reef areas worldwide, and they can be easily compared as we showed above. In the reef area close to the touristic sites, we see the number of individuals are small, but in non-reef areas, we do not have individuals at all, and opportunists are dominant in coastal stations where people step on reefs!! Coastal, deeper and in the south stations have *Amphistegina gibbosa* with symbiont bearer foraminifer's species dominating while *Amphisorus* is dominant in the inner reef areas.

**Keywords** Sedimentary reef · Diversity · Propagules · Ecology · Succession · Community · Stable · Quantitative · Spatial

## Introduction

Before talking about "Biodiversity" we should explain what "Diversity" is. The idea of diversity drives cultural, economic, social, and environment discussions all over our Planet Earth. Research shows "diversity" is strongly associated with human race and racial diversity, but, it is only one of the dimensions in our reality, because we are different in gender, language, habits, culture, social roles, differences in brain function, sexual orientation, education, skills, income, buying power, and countless others. Research shows also that differences actually make it harder for people to connect and empathize with each other, while similarities unite people.

In an ecological scope, similarities of the sediment-water interface properties of the marine environment attracts similar specific indicator organisms, which dominate. On the other hand, the condition of more diverse environments availability to be colonized will then be composed of differing species with differing characteristics. This is "diversity". Diversity has many synonyms such as assortment, diverseness, heterogeneity, heterogeneousness, manifoldness, miscellaneousness, multifariousness, multiplicity, variety, variousness, and all indicating differences!

An ecological community in which populations of plants or animals remain stable and exist in balance with each other and their environment is known as a

"Community in Climax", as its diversity is at its maximum. This is the final stage of succession, which would have remained for centuries and millennia, relatively unchanged until recently in the modern age in the Anthropocene. In this new Era, the Planet Earth is being destroyed by human based events such acute industrial accidents or chronic domestic or industrial human interference through effluents discharge daily in already threatened fresh and saline waters of rivers, bays and estuaries, coastal and deep marine waters.

In relation to diversity and their synonyms, large-scale distributions of plants and animals have been extensively studied to form concepts and paradigms in ecology. We have to recognize that larger Benthic Foraminiferal association with symbiotic organisms, and their sensitivity to changes in abiotic parameters of surrounded water and sediment, in climax ecosystems have the ability to propagate diversity adjacent to, and in near areas. Areas better able to face environmental changes movement of currents and bioturbation increase biodiversity. We infer that higher biodiversity from a climax ecosystem can propagate and increase diversity by exporting propagules to surrounding and adjacent marine areas.

We will show the quantitative assessment of how spatial and environmental variables shape the biodiversity of Larger Forams with their symbiotic organisms in two sedimentary reef areas (Pirangi and Maracajau) and our recommendation to implement preservation areas in one of them where propagules emission are capable of reaching other adjacent marine sites improving the diversity of neighboring areas. If one very biodiverse part of the whole is totally preserved over time, it actually can benefit the whole by increasing the total diversity of these two differentiate ecosystems examples.

## *Sedimentary Reefs*

Reefs dominated either by corals or by sedimentary algae are declining globally as climate change and local water quality press environmental conditions beyond the physiological tolerances of holobionts—the collective of the host and its microbial symbionts.

The high diversity of ecosystems climax, like marine reef areas, makes them highly competitive and resistant to biological invasions. However, if the environment has lost its biodiversity it has the tendency of being invaded by exotic species. The relationship between Benthic foraminifera with symbiont as holobiont, and with no symbiont as opportunists to stress tolerance through community diversity metrics suggests that symbiont community diversity metrics may function as indicators of resilience and have potential applications in diverse disciplines from climate change adaptation, mitigation and remediation. Here we will explain how, in three different reef areas.

High species diversity in small-protected plots enhances invasion resistance by increasing crowding and species richness in localized marine neighborhoods. Both the establishment (number of invaders) and success (large proportion of invaders) of invading species are reduced. These results suggest that local biodiversity represents an important line of defense against the spread of invaders and fauna replacement, a common problem of decrease in precious biodiversity.

## *Symbiont-Bearing Species*

Numerous publications have shown that certain benthic foraminifera (species of *Amphistegina* in particular), that thrive in and around coral reef habitats, are affected by global or local environmental stresses the same way hermatypic corals are. These species belong to several families, both porcelaneous and hyaline, but they all act as hosts to diverse algal endosymbionts comparable to the zooxanthellae of corals. "The potential advantages of algal symbiosis to foraminifers lie in at least three major areas: (a) energy from photosynthesis; (b) enhancement of calcification; and (c) uptake of host metabolites by symbiotic algae" (Hallock 1999).

As with hermatypic corals, these symbiont-bearing foraminifera (SBF), including several relatively large species prefer nutrient-poor, shallow, warm-water environments (e.g., Hallock 2000). Using this core idea, but expanding the inquiry to include the entire reef community of benthic foraminifera, Hallock et al. (2003) developed a proxy water-quality measure (relevant to coral-reef health) called FORAM Index (FI); the concept was further clarified by Hallock (2012). The computation of FI is based on the proportions of three species groups: symbiont bearing, "opportunistic" or "stress-tolerant" (nomenclature changed in Carnahan et al. 2009), and other small taxa.

The FORAM Index has been successfully used in coral-reef environmental studies (related to nutrification) in diverse regions, e.g., off the coasts of Florida (Hallock et al. 2003; Carnahan et al. 2009), Puerto Rico (Hallock et al. 2003), Colombia (Velásquez et al. 2011), Australia (Schueth and Frank 2008; Uthicke and Nobes 2008; Narayan and Pandolfi 2010; Uthicke et al. 2010), and Kiritimati/Christmas Island (Carilli and Walsh 2012). However, two studies from Indonesia (Natsir and Subkhan 2011) and Fiji (Strotz and Brock 2006) do not corroborate the relationship between higher FI values and coral growth and lower nitrification. In the first case, the trend is unclear; in the second, the relation is negative.

In Brazil, FI has been computed in several studies of reef foraminifera off the state of Bahia; the results, overall, do not fit the model established by Hallock et al. (2003) in southern Gulf of Mexico and northern Caribbean Sea. In both nearshore and offshore reef areas of Bahia (Corumbau and Abrolhos), relatively low FI values are widespread, and suggest a deterioration of water quality, although coral communities may be thriving at many sampling sites; in contrast, high FI values were found at some sites of low coral cover but high macroalgal cover (Barbosa et al. 2009). Oliveira-Silva et al. (2012) and Barbosa et al. (2012) also found such anomalies in further studies of Abrolhos reefs. They suggested, "Palimpsest sediment can disguise FI so that, in some cases, the index fails to correlate with coral cover" (Barbosa et al. 2012). An investigation of historical FI trends in northern Bahia

(Kelmo and Hallock 2013) was not directed toward assessment of coral-reef health, but it demonstrated the usefulness of the index in tracking environmental changes related to El Nino Southern Oscillation (ENSO) events.

This chapter will show higher biodiversity of Larger Forams with their symbiotic organisms as means of propagules emission in two sedimentary Pirangi and Maracajau reef areas (in the Rio Grande do Norte state) based on sampled abiotic and ecological indexes from sediment and water. We also recommend that the implementation of a specific marine preservation area in Pirangi reef is mandatory where biodiversity is so high that propagules emission at this site is capable of reaching other adjacent marine sites to improve the diversity of neighboring areas.

## Field and Analysis

### *Sampling Sites*

Surface sediment was scooped by scuba divers in 55 stations in Pirangi and in 40 stations in Maracajaú Reef areas in June, 2013 and July, 2014 (Figs. 9.1 and 9.2). Sampling covered reef areas, sandy sediments, and macroalgae substratum. The same areas were sampled twice, but exact station locations differed from 2013 to 2014.

### *Marine Sediment Cores*

Figure 9.3 shows the marine sediment cores that were collected in the Pirangi area. We will show you in this chapter data of one of the cores (T1).

### *Laboratory Methods*

Recent sediment samples was recovered with a Van Veen grab sampler and the uppermost 1 cm was subsampled for foraminiferal analysis. Samples were stored in a mixture of 1 g rose Bengal in distilled water to identify live specimens. Processing of sediments followed standard procedures (Eichler et al. 2007) where a fixed volume of 50 $cm^3$ of sediment was washed through a 0.063-mm sieve. After drying, samples were split using a microsplitter into subsamples of 100 living specimens. All specimens were counted when the number of foraminifers in a sample was less than 100. Species identification and counting of dry specimens were done under an optical microscope. Only Rose Bengal was used to stain a few tests, and all our data are based on total Forams counts, and scanning electron micrographs were obtained

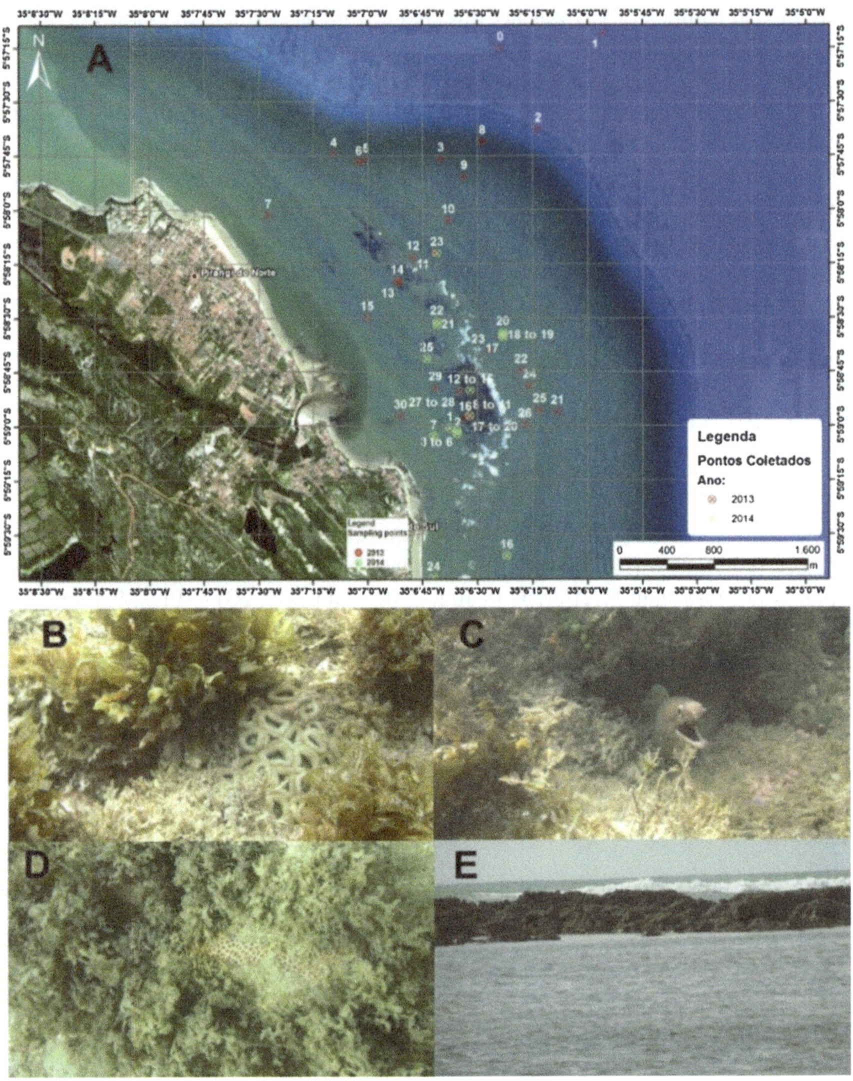


**Fig. 9.1** (**a**) Map with Pirangi Reefs sampling sites (2013 and 2014). (**b–d**) are coral reef submerse images. (**e**) A view of Pirangi Reef area

to clarify ambiguous identifications. Field recording of water temperature, salinity, and oxygen were performed, but, within either area, they did not show any noticeable variations or trends. Size fractions of sediments were also analyzed, and compared with Forams in a shallow environment of less than 15 m depth.

Marine sediment cores were taken manually, and after drying it for several weeks, they were cut in half, and pictures were taken. After that, we have sub sampled Forams fauna every 2 cm. Cores length varied from 20 to 25 cm and based on

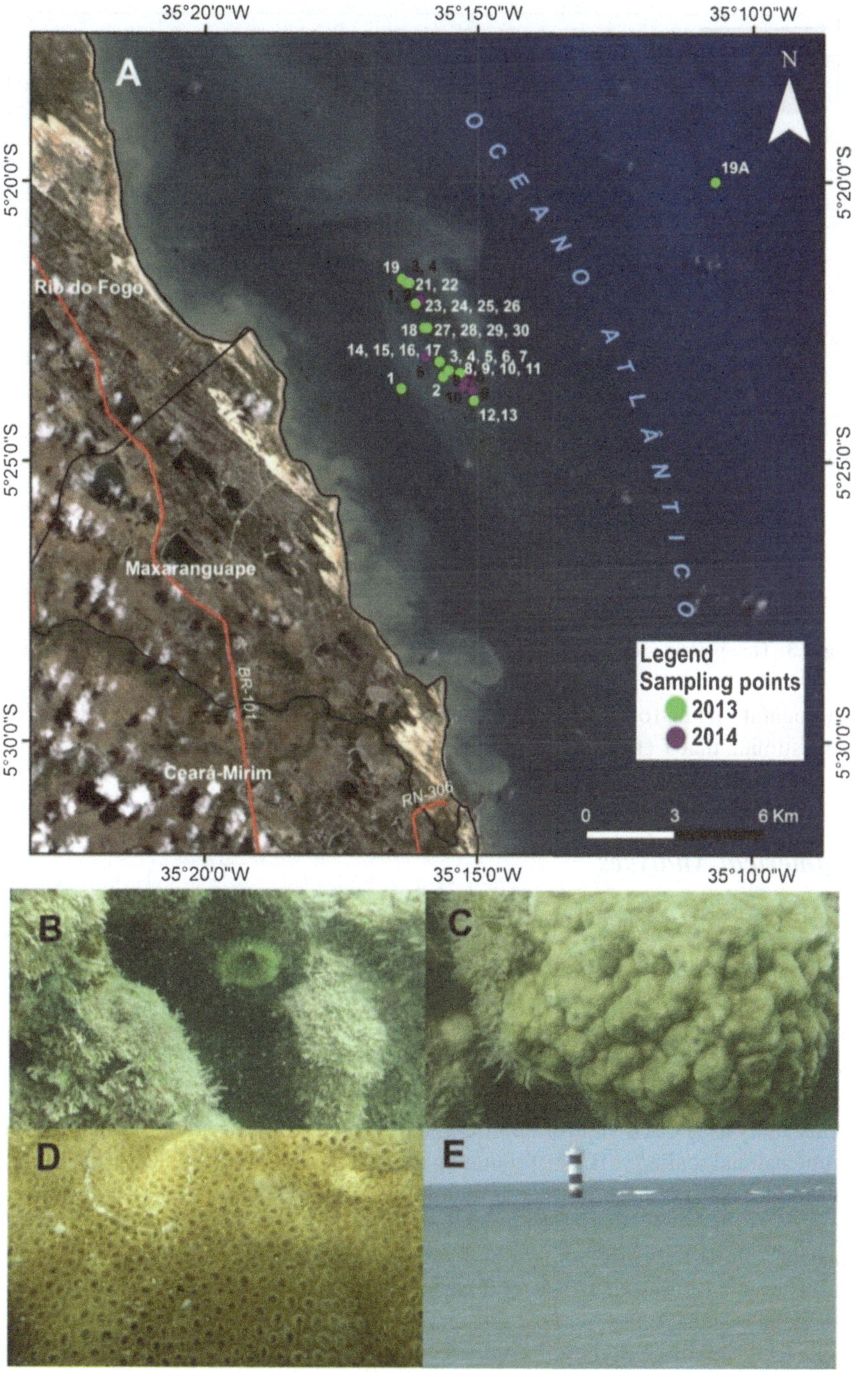


**Fig. 9.2** (**a**) Map with Maracajaú Reefs sampling sites (2013 and 2014). **b**, **c**, **d** are coral reef submerse images. (**e**) A view of Maracajaú Reef area

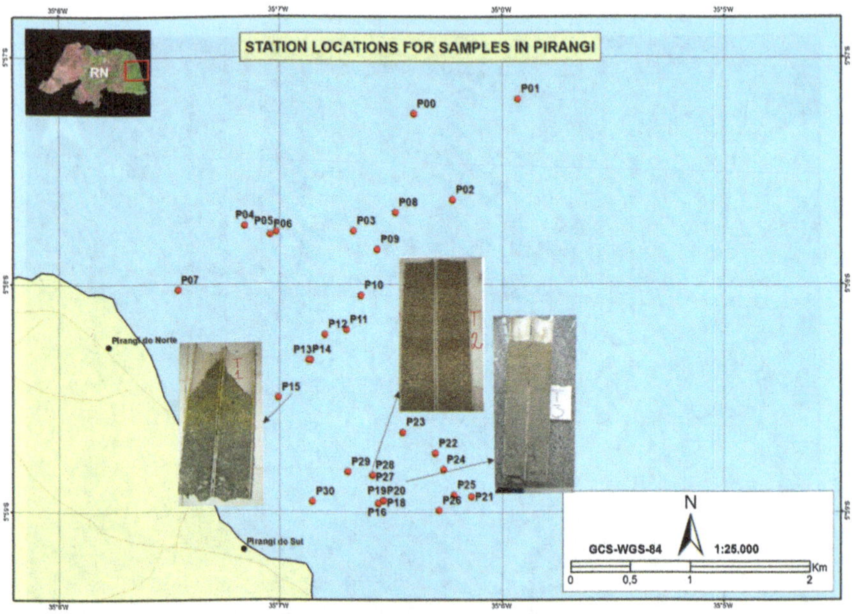


**Fig. 9.3** Three marine sediment cores (T1, T2 and T3) that were collected in Pirangi reef area

sedimentation rate for inner continental shelves (Patchineelam and Smoak 1999) we can estimate that 1 cm equals 1 year into the past.

## Numerical Analyses

*Univariate techniques*: Diversity and dominance indices are effective when used together to assess changes in community structure. Pielou eveness, Shannon-Wiener diversity, and Simpson dominance indexes (Zar 1984) were computed. These calculations were done using the Primer program developed at the University of Plymouth (Clarke and Warwick 1994a, b).

*Multivariate analyses*: Were applied to both environmental [principal components analysis (PCA)] and foraminiferal data [cluster analysis and non-metric multidimensional scaling (MDS)], producing a "map" of samples in which the placement of samples reflects the similarity of their biological communities and environmental patterns, rather than their simple geographical location. The BIOENV or BEST analysis allows the correlation of biotic datasets to find the best match between the multivariate patterns of the assemblages. The numerical procedures of diversity, dominance and evenness, PCA, cluster, MDS, BIOENV/ BEST were carried out using a University of Plymouth computer program PRIMER v6 (Plymouth Routines in Multivariate Ecological Research PRIMER-E Ltd., Plymouth) which is described in Clarke (1993), Clarke and Warwick (1994a, b), and Clarke and Ainsworth (1993).

The BEST analyses or Biota and Environment (BIOENV) matching were applied using Spearman Rank correlation method and resemblance measure as Euclidean distance.

*FORAM Index (FI)* (Hallock 2012; Hallock et al 2003): The computation of FI is based on the proportions of 3 species groups: "symbiont-bearing", "opportunistic" or "stress-tolerant" (nomenclature changed in Carnahan et al. 2009), and "other small taxa". If the proportions of these groups are Ps, Po, and Ph, respectively, then FI is given by $(10 \times Ps) + Po + (2 \times Ph)$, and an FI value of >4 indicates an environment "conducive to reef growth." The index is heavily weighted in favor of the symbiont-bearers; "to have an FI > 2, there must be some symbiont-bearing taxa, and for FI > 4, symbiont-bearing taxa must make up at least 25% of the assemblage" (Carnahan et al. 2009). The symbiont-bearing foraminifera in Rio Grande do Norte (RN) reefs are seven species: 1. *Amphistegina gibbosa, 2. Amphisorus hemprichii, 3. Archaias angulatus, 4. Borelis schlumbergeri, 5. Heterostegina depressa, H. antillarum, 6. Peneroplis carinatus 7. Laevipeneroplis proteus* (Fig. 9.4).

## Results

### *Spatial Distributed Stations*

Table 9.1 from Pirangi 2013 shows depth varies from 0.7 to 13.5 m. Table 9.1 and Fig. 9.5 shows data and illustration of size fractions of sediment indicate higher percentage of coarser fraction at stations 17 and 27 followed by stations 1 and 3. The lowest value of coarse fraction was found at stations 7, 15, 30, 10 and 22, and highest percentage of sand was found at stations 2, 8, 10, 15, 22, and 30. Most are outer reef, with exception of coastal station 30. The lowest value for sand was found 3, 17 and 27. Silt and clay average was its highest on stations 3, 7, and 29. The total organic matter (TOM) was highest at stations 7 and 22. Higher $CaCO_3$ was found at 17 and 18 close to the touristic area and 27 and 28 from outer reef.

**Fig. 9.4** Symbiont-bearing foraminifera in RN reefs: 1. *Amphistegina gibbosa, 2. Amphisorus hemprichii, 3. Archaias angulatus, 4. Borelis schlumbergeri, 5. Heterostegina depressa, H. antillarum, 6. Peneroplis carinatus 7. Laevipeneroplis proteus*

**Table 9.1** Granulometry of sediment from Pirangi 2013

| Pirangi 2013 | Depth (m) | Coarse (%) | Sand (%) | Silt (96) | Clay (%) | TOM (%) | $CaCO_3$ (%) |
|---|---|---|---|---|---|---|---|
| 1 | 12 | 27.84 | 70.82 | 1.07 | 0.27 | 0.03 | 32.80 |
| 2 | 9.5 | 2.08 | 97.75 | 0.13 | 0.03 | 0.83 | 3.70 |
| 3 | 12 | 23.62 | 56.65 | 10.99 | 8.75 | 3.23 | 58.70 |
| 7 | 5.4 | 0.00 | 65.16 | 16.96 | 17.88 | 9.10 | 39.40 |
| 8 | 13 | 3.84 | 96.16 | 0.00 | 0.00 | 1.03 | 19.40 |
| 9 | 13.5 | 16.92 | 82.91 | 0.17 | 0.00 | 2.07 | 44.80 |
| 10 | 10 | 0.57 | 99.43 | 0.00 | 0.00 | 1.27 | 73.90 |
| 11 | 11 | 11.62 | 88.21 | 0.17 | 0.00 | 1.63 | 81.00 |
| 15 | 4.6 | 0.00 | 98.73 | 1.27 | 0.00 | 4.93 | 56.10 |
| 17 | 0.7 | 44.54 | 54.95 | 0.51 | 0.00 | 2.70 | 81.10 |
| 18 | 1 | 17.70 | 81.65 | 0.65 | 0.00 | 2.80 | 84.00 |
| 22 | 11 | 0.08 | 99.41 | 0.51 | 0.00 | 14.67 | 45.20 |
| 26 | 4 | 9.93 | 89.14 | 0.93 | 0.00 | 3.00 | 32.30 |
| 27 | 2.5 | 40.50 | 57.61 | 1.90 | 0.00 | 3.37 | 88.10 |
| 28 | 2.5 | 15.99 | 84.01 | 0.00 | 0.00 | 2.87 | 98.40 |
| 29 | 3.5 | 4.41 | 60.43 | 14.58 | 20.57 | 4.67 | 50.60 |
| 30 | 3.5 | 0.00 | 99.05 | 0.95 | 0.00 | 2.07 | 23.60 |

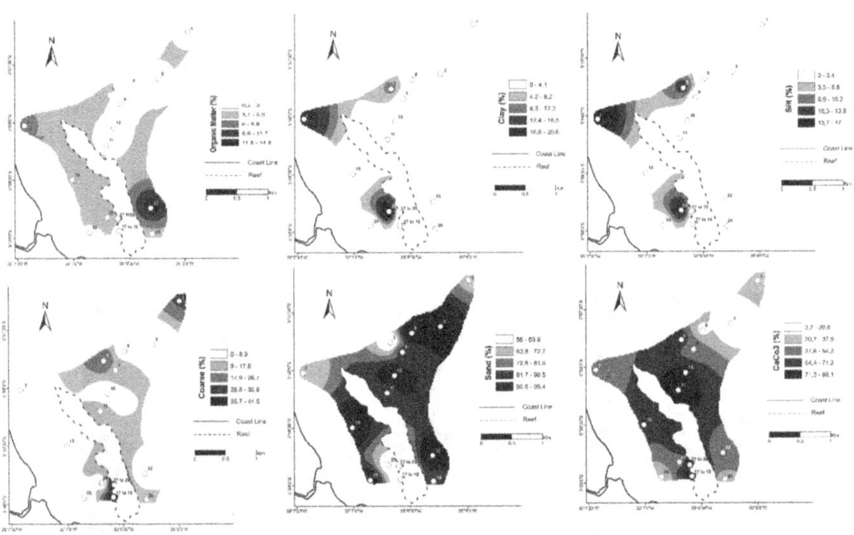

**Fig. 9.5** Percentage of organic matter, clay, silt, coarse, sand and $CaCO_3$ in Pirangi for 2013

Table 9.2 Cumulative variation of abiotic variables that explain the foraminiferal fauna

| PC | Eigen values | % Variation | Cum.%Variation |
|---|---|---|---|
| 1 | 2.6 | 37.1 | 37.1 |
| 2 | 2.14 | 30.6 | **67.7** |
| 3 | 1.11 | 15.8 | 83.5 |
| 4 | 0.723 | 10.3 | 93.8 |
| 5 | 0.41 | 5.9 | 99.7 |

## *PCA (Principal Component Analysis)*

### Pirangi 2013

The axes PC1 and PC2 explain 67.7% of the PCA correlation (Table 9.2) among variables from Pirangi 2013 (Table 9.3). PCA analysis confirmed stations 17 and 27 with highest loading of coarse fraction (Fig. 9.4) from outer reef and from touristic area. Stations 17, 27, 28, and 18 have the highest $CaCO_3$ percentage. Station 22 is the deepest with highest sand percentage. Stations 29 and 7 have the highest percentage of silt, clay and TOM.

PCA analysis show that the two first principal components (PC1 and PC2) represent up to 67.7% of the total variance in Pirangi (Table 9.2). The highest positive values for PC1 explain percentage of sand availability at those stations followed by depth (Table 9.3 and Fig. 9.6). The highest negative values of PCA explain percentage of silt and clay which also contributed to the occurrence of heterotrophic (*Bolivina, Elphidium, Nonion*) and stress-tolerant foraminifera (*Cibicides, Discorbis, Miliolinella, Pyrgo, Quinqueloculina, Rosalina,* and *Triloculina*). Coarse and calcium carbonate content, contribute to the construction of positive values of PC2.

### Maracajau 2013

Table 9.4 from Maracajau 2013 shows depth varied from 2.1 to 5.4 m. Table 9.4 and Fig. 9.7 shows data and illustration of size fractions of sediment indicate higher percentage of coarser fraction, was found at stations 17 and 24 followed by stations

Table 9.3 PCA axes values for abiotic variables

| Variable | PC1 | PC2 | PC3 | PC4 | PC5 |
|---|---|---|---|---|---|
| Coarse (%) | **−0.135** | **0.607** | −0.254 | −0.23 | −0.364 |
| Sand (%) | **0.506** | **−0.317** | 0.295 | 0.108 | 0.183 |
| Silt (%) | **−0.559** | **−0.261** | −0.126 | 0.082 | 0.136 |
| Clay (%) | **−0.544** | **−0.285** | −0.102 | 0.155 | 0.207 |
| TOM (%) | **−0.212** | **−0.303** | 0.51 | −0.698 | −0.337 |
| $CaCO_3$ (%) | **−0.147** | **0.489** | 0.405 | −0.222 | 0.725 |
| Depth (m) | **0.225** | **−0.226** | −0.63 | −0.606 | 0.365 |

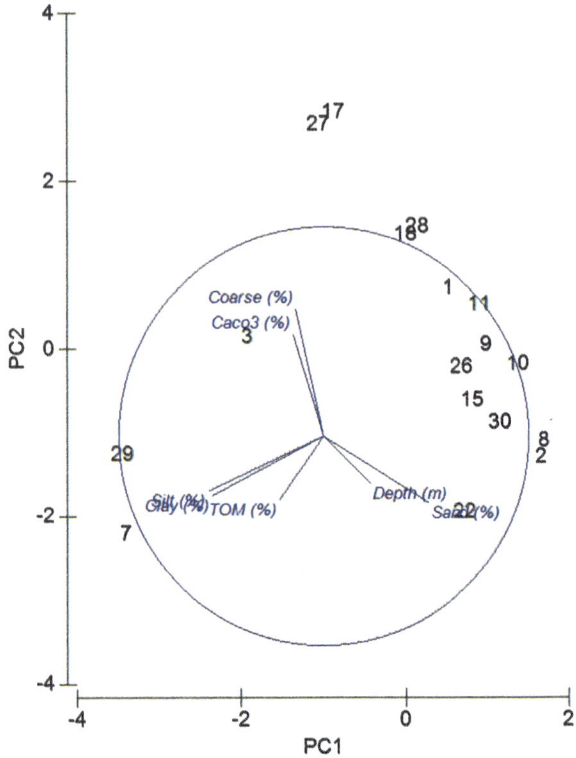


**Fig. 9.6** Plot representing PCA analysis for Pirangi 2013

20 and 22. The lowest value of coarse fraction was found at stations 2, 4, 5, and 11, and highest percentage of sand was found at stations 1, 2, 8, 12, 13, 19, and 22. The lowest value for sand was found at 5, 7, 17, and 27. Silt and clay average was its highest on stations 5, 7, 17, 23, and 27. The lowest percentage of silt and clay were found at 1, 2, 12, 13, 19, 20, 21, 22, and 24. The total organic matter (TOM) was highest at stations 7 and 22, and it is lowest at 12. Higher $CaCO_3$ was found at 17 and 18 close to the touristic area and 27 and 28 from outer reef.

PCA analysis show that the two first principal components (PC1 and PC2) represent up to 76.5 of the total variance in Maracajaú (Table 9.5).

The highest positive values for PC1 show that percentage of sand and $CaCO_3$ availability and depth at those stations (Table 9.5 and Fig. 9.8) are the most influencing factors in foraminiferal fauna. The highest negative values show that percentage of silt, clay, and TOM contribute to the occurrence of other heterotrophic (*Bolivina, Elphidium, Nonion*) and stress-tolerant foraminifera (*Cibicides, Discorbis, Miliolinella, Pyrgo, Quinqueloculina, Rosalina*, and *Triloculina*) similar to what is hapenning in Pirangi.

**Table 9.4** Granulometry of sediment from Maracajau 2013

| Maracajaú 2013 | Depth (m) | Coarse (%) | Sand (%) | Silt (%) | Clay (%) | TOM (%) | $CaCO_3$ (%) |
|---|---|---|---|---|---|---|---|
| 1 | 3.5 | 1.70 | 97.72 | 0.48 | 0.10 | 2.30 | 93.90 |
| 2 | 4 | 0.00 | 99.93 | 0.07 | 0.00 | 2.70 | 98.60 |
| 3 | 2.5 | 0.43 | 70.26 | 15.08 | 14.23 | 4.83 | 87.30 |
| 4 | 3.5 | 0.08 | 61.56 | 15.78 | 22.59 | 6.13 | 85.00 |
| 5 | 3.5 | 0.00 | 32.76 | 34.02 | 33.22 | 15.03 | 76.40 |
| 6 | 3.5 | 2.06 | 44.60 | 22.66 | 30.67 | 6.40 | 84.90 |
| 7 | 3.5 | 1.06 | 37.59 | 32.98 | 28.37 | 4.57 | 86.60 |
| 8 | 2.5 | 1.61 | 96.14 | 1.95 | 0.31 | 3.23 | 96.20 |
| 9 | 2.5 | 4.16 | 60.24 | 19.39 | 16.21 | 8.18 | 88.00 |
| 10 | 2.5 | 4.84 | 87.11 | 6.50 | 1.56 | 3.83 | 97.90 |
| 11 | 2.5 | 0.14 | 79.07 | 5.48 | 15.31 | 2.63 | 97.10 |
| 12 | 4 | 2.86 | 96.80 | 0.27 | 0.07 | 0.97 | 99.10 |
| 13 | 3.2 | 1.05 | 98.95 | 0.00 | 0.00 | 3.30 | 97.70 |
| 14 | 2.1 | 2.73 | 72.83 | 10.93 | 13.50 | 4.53 | 76.60 |
| 15 | 3.7 | 1.93 | 51.24 | 29.57 | 17.26 | 4.70 | 92.50 |
| 17 | 2.5 | 20.24 | 36.86 | 35.66 | 7.24 | 7.67 | 63.67 |
| 18 | 2.5 | 1.96 | 75.72 | 15.06 | 7.26 | 3.47 | 97.20 |
| 19 | 4 | 8.39 | 91.10 | 0.41 | 0.10 | 3.40 | 99.40 |
| 20 | 4 | 10.92 | 88.02 | 0.85 | 0.20 | 2.40 | 97.70 |
| 21 | 5 | 7.31 | 92.70 | 0.00 | 0.00 | 2.93 | 97.50 |
| 22 | 5.4 | 8.12 | 91.17 | 0.58 | 0.14 | 2.37 | 90.80 |
| 23 | 3.9 | 0.66 | 28.86 | 35.77 | 34.71 | 7.27 | 79.10 |
| 24 | 3.9 | 15.76 | 82.99 | 1.08 | 0.17 | 4.27 | 98.50 |
| 25 | 3.9 | 6.82 | 84.34 | 7.10 | 1.73 | 3.93 | 75.60 |
| 27 | 3.5 | 0.99 | 35.04 | 31.53 | 32.44 | 6.40 | 82.40 |
| 28 | 3.5 | 1.82 | 49.67 | 22.20 | 26.31 | 4.87 | 88.40 |

# BEST

The BEST analyses or Biota and Environment (BIOENV) reveal coarse fraction as the best-correlated variable with Forams abundance followed by sand in both Pirangi and Maracajau reefs in 2013. List of best-correlated variable from BIOENV for both Pirangi and Maracajau reefs (2013 and 2014): 1 Coarse (%) 2 Sand (%) 3 Silt (%) 4 Clay (%) 5 TOM (%) 6 $CaCO_3$ (%) 7 Depth (m).

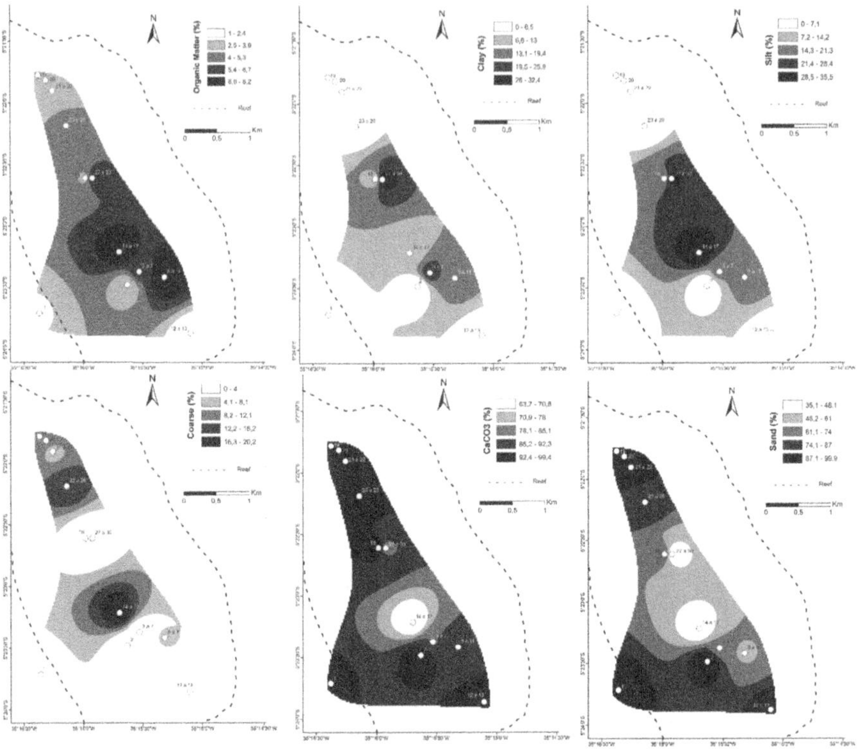


**Fig. 9.7** Percentage of organic matter, clay, silt, coarse, sand and $CaCO_3$ in Maracajau for 2013

## *Forams Ecology*

Our study showed that based on number of species (S) and number of individuals (N) ecological indexes Evenness (J′), diversity (H′(loge), and dominance (1−Lambda′) were able to show that in general, samples from Maracajaú were more diverse than Pirangi in both years (Tables 9.6 and 9.7). The dominant species in both reef areas is *Quinqueloculina lamarckiana* followed by *Amphistegina gibbosa*.

**Table 9.5** PCA axes values for abiotic variables and cumulative variation of abiotic variables that explain the foraminiferal fauna

| PC | Eigen values | %Variation | Cum.%Variation |
|---|---|---|---|
| 1 | 4.12 | 58.8 | 58.8 |
| 2 | 1.24 | 17.7 | **76.5** |
| 3 | 0.936 | 13.4 | 89.8 |
| 4 | 0.34 | 4.9 | 94.7 |
| 5 | 0.296 | 4.2 | 98.9 |

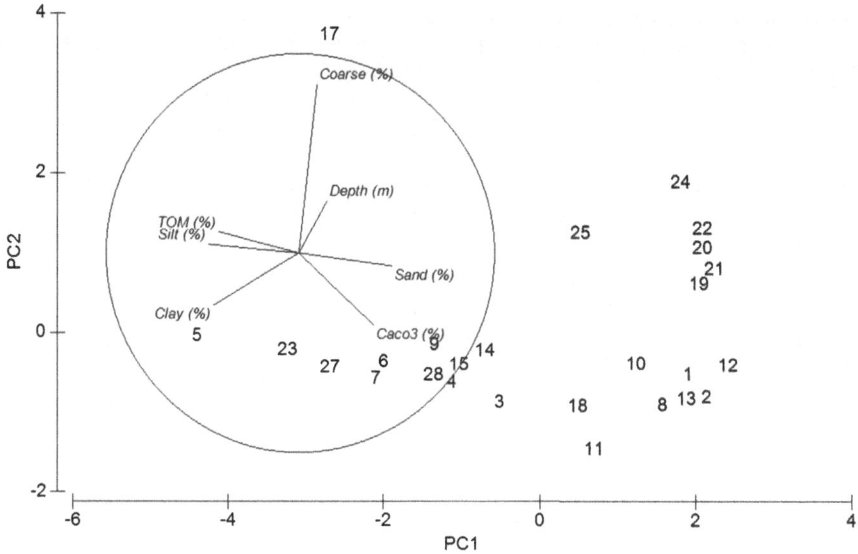


**Fig. 9.8** Plot representing PCA analysis for Maracajau 2013

## Cluster Analysis

This analysis shows in Fig. 9.9 that Pirangi 2013 were divided in three groups: Group I: 2 and 28; Group II: 8 and 11 is composed by stations with less number of individuals and no SBS. Group III: 30, 21, 12, 26, 1, 10, 23, 3, 9, 22, 24, 25 formed by coastal stations, medium diversity, with most stations containing greater proportion of SBS and Group IV: 27, 16, 7, 15, 17, 20, 18, 29 composed by deeper stations with higher number of species this group contains the stations (17 and 18) that *Amphisorus hemprichii* has higher proportions. In Pirangi 2014, we see the formation of four groups: Group I is formed by stations: 12, 15, 22, Group II: 21, 11, 20, 7, 2, 6, 24, 25, 16, 19, 17, 18; Group III: 3, 5, 10, 8, 9, Group IV is composed by 13, 14, 23. Stations 1 and 4 stand itself without forming any groups.

In Maracajau 2013, five groups were formed: Groups I: 6, 26; Group II; 5, 7, 17, Both groups with only *Amphisorus hemprichii*. Group III: 16, 4, 10, 15, 11, 9, 14, 3, 8, with station with no individuals to few ones; and Group IV: 23, 25, 28, 27, 30, 22, 29 with the higher diverse stations and most of them with *Amphisorus hemprichii* and Group V: 4, 18, 9, 10, 20, 1, 12, 13, 2, 19, 21, 18, 24 with *Amphistegina* dominating less diverse stations. In Maracajau 2014, two groups were formed: Group I: 1, 2, 5, 6 with more diverse stations and the presence of *Amphisorus hemprichii* and Group II: 4, 8, 9, 10 with less diverse stations and the dominance of *Amphistegina*, and station 3 stands by itself because there is just one species (*Quinqueloculina lamarckiana*) living in this area.

**Tables 9.6 and 9.7** Number of species (S), number of individuals (N) Evenness (J′), diversity (H′(loge) and dominance (1−Lambda′) of Forams species from Pirangi and Maracajaú (2013 and 2014)

| | S | N | J′ | H′(loge) | 1−Lambda′ |
|---|---|---|---|---|---|
| Pirangi 2013 | | | | | |
| 1 | 8 | 212 | 0.52 | 1.08 | 0.59 |
| 2 | 2 | 8 | 1.00 | 0.69 | 0.57 |
| 3 | 8 | 294 | 0.48 | 1.01 | 0.57 |
| 7 | 15 | 270 | 0.60 | 1.63 | 0.69 |
| 8 | 2 | 228 | 1.00 | 0.69 | 0.50 |
| 9 | 6 | 228 | 0.50 | 0.90 | 0.55 |
| 10 | 6 | 194 | 0.53 | 0.94 | 0.56 |
| 11 | 3 | 62 | 0.81 | 0.89 | 0.57 |
| 12 | 8 | 94 | 0.65 | 1.35 | 0.66 |
| 15 | 15 | 208 | 0.58 | 1.56 | 0.67 |
| 16 | 16 | 162 | 0.66 | 1.83 | 0.72 |
| 17 | 7 | 118 | 0.73 | 1.43 | 0.68 |
| 18 | 13 | 132 | 0.68 | 1.75 | 0.72 |
| 20 | 12 | 72 | 0.71 | 1.75 | 0.72 |
| 21 | 8 | 176 | 0.63 | 1.31 | 0.64 |
| 22 | 5 | 216 | 0.63 | 1.01 | 0.59 |
| 23 | 6 | 438 | 0.55 | 0.98 | 0.58 |
| 24 | 5 | 244 | 0.65 | 1.05 | 0.60 |
| 25 | 5 | 208 | 0.63 | 1.02 | 0.58 |
| 26 | 10 | 440 | 0.48 | 1.11 | 0.59 |
| 27 | 13 | 58 | 0.69 | 1.76 | 0.72 |
| 28 | 5 | 16 | 0.82 | 1.32 | 0.72 |
| 29 | 13 | 110 | 0.67 | 1.72 | 0.71 |
| 30 | 11 | 244 | 0.44 | 1.06 | 0.58 |
| Pirangi 2014 | | | | | |
| 1 | 9 | 158 | 0.55 | 1.20 | 0.61 |
| 2 | 9 | 176 | 0.61 | 1.34 | 0.65 |
| 3 | 12 | 274 | 0.59 | 1.48 | 0.68 |
| 4 | 14 | 142 | 0.64 | 1.69 | 0.70 |
| 5 | 10 | 204 | 0.61 | 1.40 | 0.67 |
| 6 | 9 | 196 | 0.61 | 1.35 | 0.65 |
| 7 | 7 | 248 | 0.60 | 1.16 | 0.61 |
| 8 | 13 | 216 | 0.55 | 1.42 | 0.66 |
| 9 | 11 | 158 | 0.60 | 1.43 | 0.66 |
| 10 | 10 | 150 | 0.60 | 1.38 | 0.67 |
| 11 | 7 | 114 | 0.64 | 1.24 | 0.64 |
| 12 | 6 | 94 | 0.69 | 1.24 | 0.65 |
| 13 | 5 | 18 | 0.79 | 1.27 | 0.69 |
| 14 | 3 | 18 | 0.87 | 0.96 | 0.62 |

(continued)

**Tables 9.6 and 9.7** (continued)

| | S | N | J′ | H′(loge) | 1−Lambda′ |
|---|---|---|---|---|---|
| 15 | 9 | 146 | 0.68 | 1.50 | 0.68 |
| 16 | 6 | 236 | 0.59 | 1.06 | 0.60 |
| 17 | 5 | 254 | 0.64 | 1.04 | 0.59 |
| 18 | 5 | 252 | 0.65 | 1.04 | 0.59 |
| 19 | 7 | 276 | 0.57 | 1.10 | 0.61 |
| 20 | 8 | 182 | 0.52 | 1.09 | 0.59 |
| 21 | 9 | 114 | 0.59 | 1.30 | 0.64 |
| 22 | 12 | 160 | 0.62 | 1.54 | 0.68 |
| 23 | 5 | 60 | 0.76 | 1.23 | 0.66 |
| 24 | 6 | 194 | 0.62 | 1.11 | 0.63 |
| 25 | 5 | 126 | 0.69 | 1.11 | 0.63 |

| | S | N | J′ | H′(loge) | 1−Lambda′ |
|---|---|---|---|---|---|
| Maracajau 2013 | | | | | |
| 1 | 12 | 212 | 0.68 | 1.70 | 0.75 |
| 2 | 12 | 202 | 0.65 | 1.60 | 0.74 |
| 3 | 19 | 162 | 0.68 | 2.02 | 0.78 |
| 4 | 9 | 106 | 0.73 | 1.60 | 0.75 |
| 5 | 7 | 22 | 0.84 | 1.64 | 0.79 |
| 6 | 10 | 30 | 0.81 | 1.86 | 0.79 |
| 7 | 10 | 36 | 0.77 | 1.78 | 0.76 |
| 8 | 19 | 178 | 0.66 | 1.93 | 0.76 |
| 9 | 19 | 190 | 0.69 | 2.02 | 0.78 |
| 10 | 19 | 160 | 0.70 | 2.07 | 0.74 |
| 11 | 19 | 94 | 0.69 | 2.02 | 0.74 |
| 12 | 9 | 350 | 0.70 | 1.54 | 0.73 |
| 13 | 10 | 260 | 0.67 | 1.54 | 0.74 |
| 14 | 18 | 170 | 0.71 | 2.06 | 0.79 |
| 15 | 19 | 142 | 0.67 | 1.96 | 0.74 |
| 16 | 12 | 50 | 0.73 | 1.80 | 0.74 |
| 17 | 10 | 36 | 0.77 | 1.77 | 0.76 |
| 18 | 17 | 228 | 0.64 | 1.81 | 0.75 |
| 19 | 11 | 252 | 0.64 | 1.54 | 0.71 |
| 20 | 10 | 60 | 0.63 | 1.46 | 0.68 |
| 21 | 21 | 294 | 0.51 | 1.56 | 0.66 |
| 22 | 13 | 80 | 0.73 | 1.87 | 0.74 |
| 23 | 30 | 408 | 0.65 | 2.20 | 0.76 |
| 24 | 25 | 258 | 0.60 | 1.94 | 0.76 |
| 25 | 22 | 194 | 0.65 | 2.01 | 0.74 |
| 26 | 10 | 22 | 0.82 | 1.88 | 0.81 |
| 27 | 22 | 168 | 0.67 | 2.08 | 0.74 |
| 28 | 21 | 182 | 0.67 | 2.03 | 0.74 |

(continued)

**Tables 9.6 and 9.7** (continued)

| | S | N | J′ | H′(loge) | 1−Lambda′ |
|---|---|---|---|---|---|
| 29 | 14 | 70 | 0.73 | 1.94 | 0.76 |
| 30 | 19 | 172 | 0.67 | 1.98 | 0.74 |
| Maracajau 2014 | | | | | |
| 1 | 17 | 238 | 0.58 | 1.65 | 0.72 |
| 2 | 16 | 208 | 0.70 | 1.94 | 0.77 |
| 3 | 3 | 128 | 0.90 | 0.98 | 0.61 |
| 4 | 7 | 130 | 0.72 | 1.40 | 0.69 |
| 5 | 13 | 202 | 0.73 | 1.87 | 0.75 |
| 6 | 17 | 228 | 0.68 | 1.94 | 0.75 |
| 8 | 11 | 226 | 0.75 | 1.79 | 0.76 |
| 9 | 9 | 256 | 0.73 | 1.61 | 0.74 |
| 10 | 10 | 238 | 0.68 | 1.58 | 0.73 |

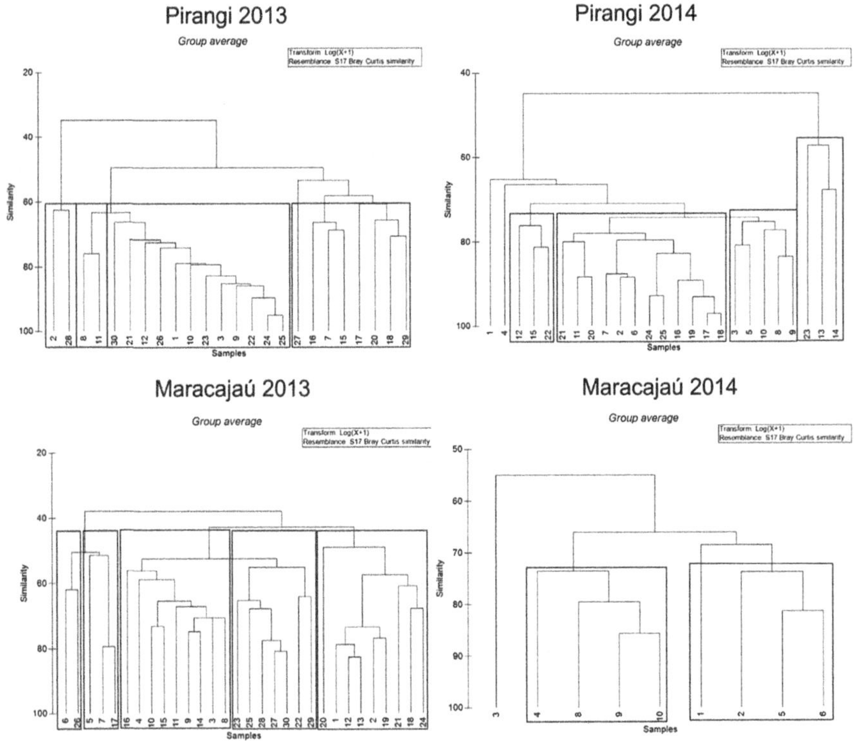


**Fig. 9.9** Cluster analysis plots of Pirangi and Maracajau for 2013 and 2014 sampling

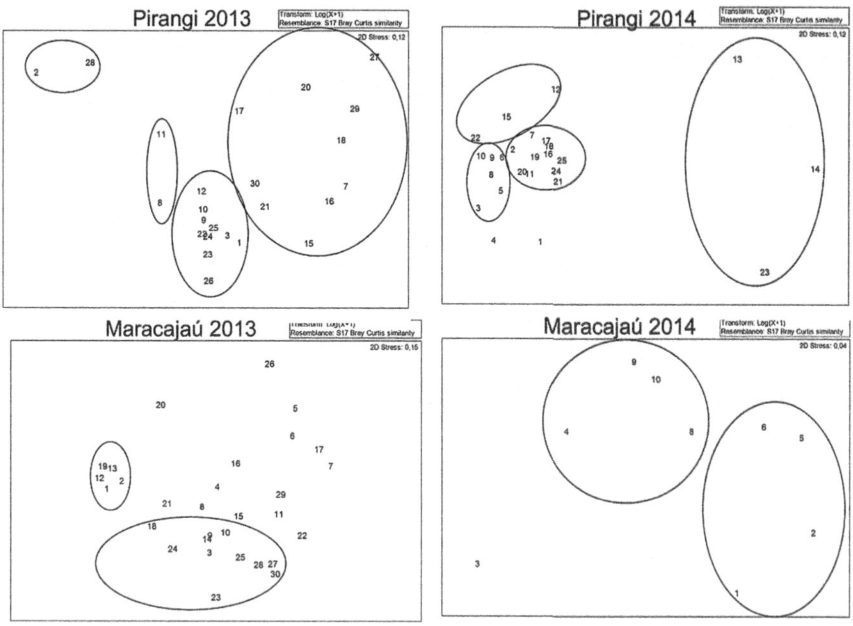


**Fig. 9.10** MDS plots of Pirangi and Maracajau for 2013 and 2014 sampling

## MDS Analysis

MDS analysis revealed that stations were differentiated by its ecological indexes (Fig. 9.10). In Pirangi, 2013, MDS analysis reveal 4 groups which includes: two groups: (I) 28, 2, 11; and (II) 1, 10, 9, 3, 21, 26, 23, 25, 24, 22, 12, 8, with deeper stations, less diverse, more dominance of one or two species; Group (III) 20, 16, 17, 18, 7, 29, 27 from reef formations, and Group (IV) 7, 15, 29, 30 from coastal stations with higher diversity. Pirangi 2014 showed that Group (I) 1, 3, 4, 5, 8, 20, 22, 10, 11, 15, 21, 24, 2, 6, 9, 7, 19, 7, 16, 17, 18, 25, 16, 12 has higher diversity; and stations 13, 14, 23 are outliers with low diversity.

Maracajau 2013 Group I (1, 2, 12, 13, 19) are more diverse stations, Group II (3, 4, 9, 14, 23, 8, 18, 21, 24), Group III (27, 30, 28, 25, 5, 6, 7, 17, 16, 22, 29, 10, 15, 11) with less diverse stations. Maracajau 2014 Group I with less diverse stations 4, 8, 9, 10; and Group II with 1, 2, 5, 6, more diverse stations, and station 3 with only one species.

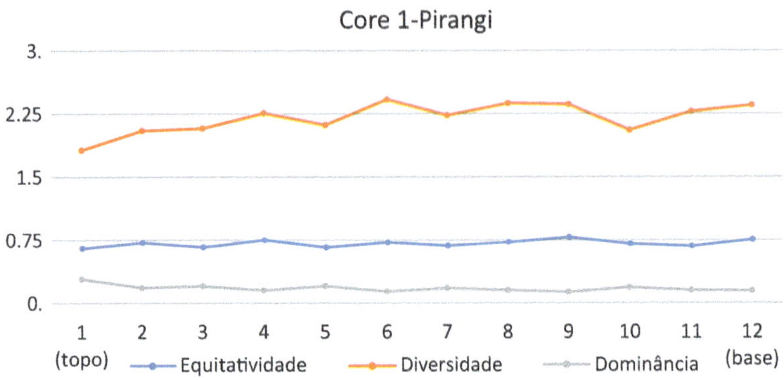


**Fig. 9.11** Evenness, diversity and dominance in Core 1

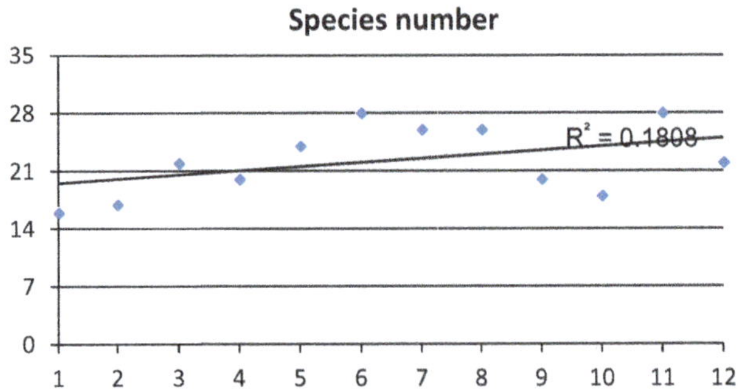


**Fig. 9.12** Species number of foraminifera from Core 1

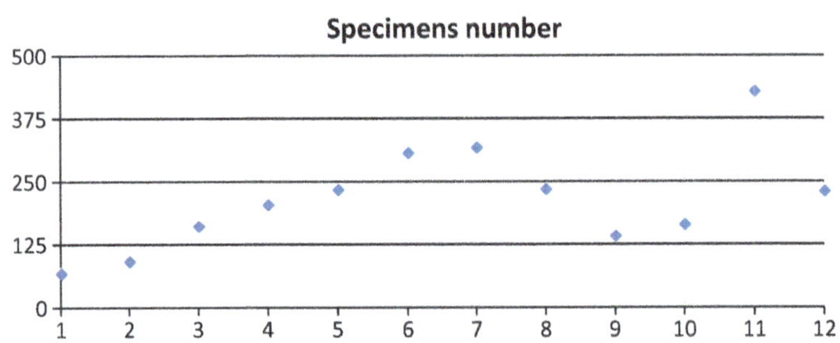


**Fig. 9.13** Specimens number of foraminifera from Core 1

## *Temporal Sampling and Paleo Results (Ecological Indexes in Cores)*

Figures 9.10 and 9.11 shows that station 10 has the lowest diversity and higher dominance. Station 6 show the highest diversity of core 1. Stations 4 and 9 show the

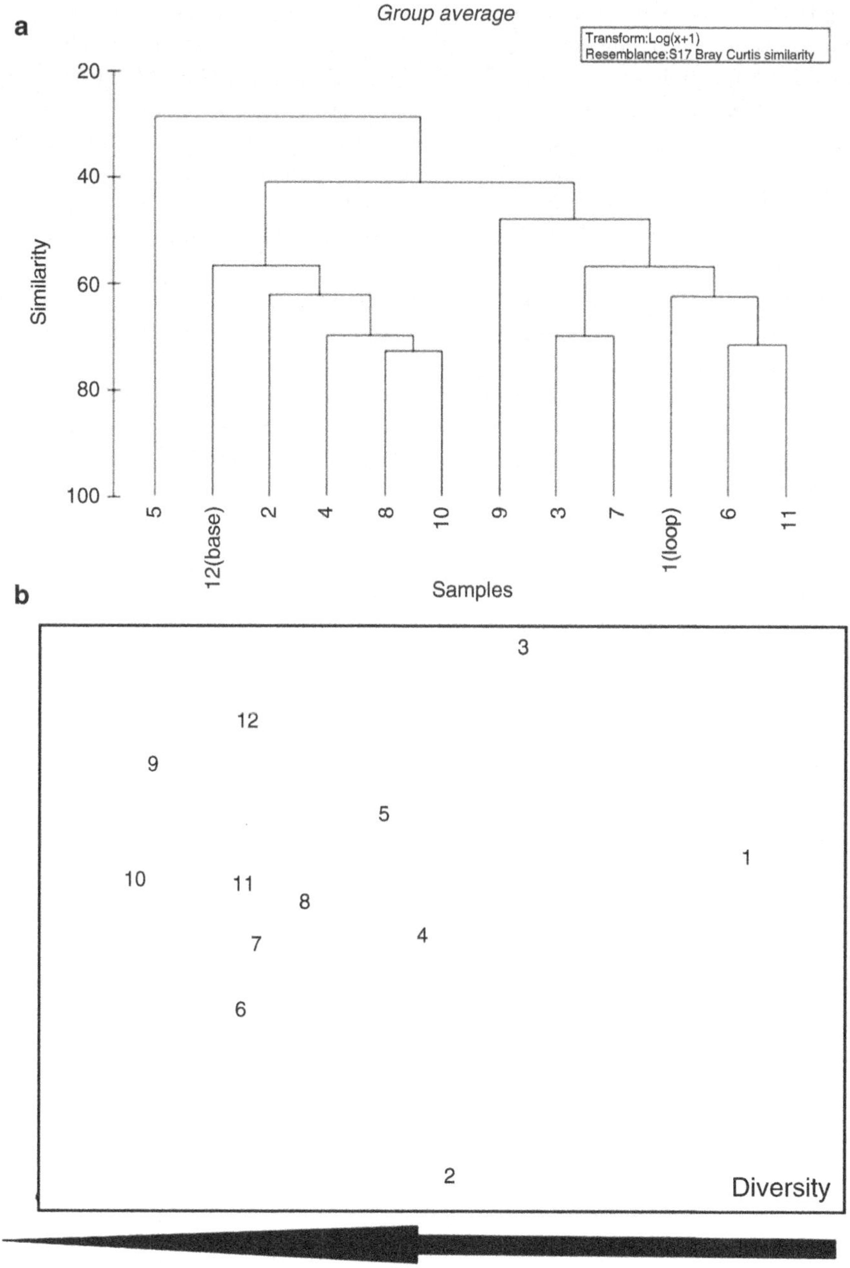


**Fig. 9.14** (**a**) Cluster analysis. (**b**) MDS analysis. Both analysis were performed on the Forams species. (**b**) shows that station 1 was the less diverse followed by station 3 and 2

highest evenness with low dominance values, which demonstrates no species dominating over others. Number of species vary from 16 to 29 and the number of species vary from 68 to 430 specimens. Evenness has less value (0.66) in station 3 and higher value (0.79) in station 9. The lowest diversity was in station 1 (1.83) and higher in station 6 (2.43). Dominance presented the lowest value (0.12) in station 9 and the highest value is in station 1 (0.28). All these ecological values help us understand how the community have evolved through time).

Number of species (Fig. 9.12) show stations that are more diverse are 6 and 11, and station 1 exhibits the lowest diversity.

Figure 9.13 shows the number of specimens from each station, revealing that station 11 has the highest number of individuals followed by stations 6 and 7. Again, station 1 has the lowest number of specimens.

*Ammonia tepida, Bolivina striatula, Discorbis peruvianus, Elphidium discoidale, Quinqueloculina lamarckiana, Q. patagonica,* and *Textularia earlandi* were present in all periods from the most recent (last 25 years) to the most ancient. Data indicate increase in diversity toward the past in every analyzed core (Fig. 9.14). Here we are just showing the data from Core 1, however every other core showed the same tendency.

Based on sedimentation rate of inner continental shelves (Patchineelam and Smoak 1999) we can estimate that 1 cm is equal 1 year in the past. Approximately 24 years ago (Core 1 was 24 cm), the environment had higher diversity and evenness, and lower dominance. We can see inter annual oscillations; however, there is

| Pirangi 2013 | 1 | Proportions | FI | 2 | Proportions | FI | 3 | Proportions | FI | 7 | Proportions | FI | 8 | Proportions | FI | 9 | Proportions | FI | 10 | Proportions | FI | 11 | Proportions | FI |
|---|---|---|---|---|---|---|---|---|---|---|---|---|---|---|---|---|---|---|---|---|---|---|---|---|
| SBS | 7 | 0,07 | 0,66 | 0 | 0,00 | 0,00 | 12 | 0,08 | 0,82 | 10 | 0,07 | 0,74 | 0 | 0,00 | 0,00 | 7 | 0,06 | 0,61 | 1 | 0,01 | 0,10 | 0 | 0,00 | 0,00 |
| Oportunists | 3 | 0,03 | 0,03 | 0 | 0,00 | 0,00 | 3 | 0,02 | 0,02 | 10 | 0,07 | 0,07 | 0 | 0,00 | 0,00 | 1 | 0,01 | 0,01 | 1 | 0,01 | 0,01 | 0 | 0,00 | 0,00 |
| small taxa | 96 | 0,91 | 1,81 | 4 | 1,00 | 2,00 | 132 | 0,90 | 1,80 | 115 | 0,85 | 1,70 | 114 | 1,00 | 2,00 | 106 | 0,93 | 1,86 | 95 | 0,98 | 1,96 | 31 | 1,00 | 2,00 |
| Total | 106 | 1,00 | 2,50 | 4 | 1,00 | 2,00 | 147 | 1,00 | 2,63 | 135 | 1,00 | 2,52 | 114 | 1,00 | 2,00 | 114 | 1,00 | 2,48 | 97 | 1,00 | 2,07 | 31 | 1,00 | 2,00 |

| 12 | Proportions | FI | 14 | Proportions | FI | 15 | Proportions | FI | 16 | Proportions | FI | 17 | Proportions | FI | 18 | Proportions | FI | 20 | Proportions | FI | 21 | Proportions | FI | 22 | Proportions | FI |
|---|---|---|---|---|---|---|---|---|---|---|---|---|---|---|---|---|---|---|---|---|---|---|---|---|---|---|
| 4 | 0,09 | 0,89 | 0 | 0,00 | | 10 | 0,10 | 0,96 | 8 | 0,10 | 0,99 | 12 | 0,22 | 2,18 | 8 | 0,12 | 1,21 | 3 | 0,08 | 0,83 | 3 | 0,03 | 0,34 | 17 | 0,16 | 1,57 |
| 3 | 0,07 | 0,07 | 0 | 0,00 | | 6 | 0,06 | 0,06 | 9 | 0,11 | 0,11 | 0 | 0,00 | 0,00 | 1 | 0,02 | 0,02 | 4 | 0,11 | 0,11 | 5 | 0,06 | 0,06 | 0 | 0,00 | 0,00 |
| 38 | 0,84 | 1,69 | 0 | 0,00 | | 88 | 0,85 | 1,69 | 64 | 0,79 | 1,58 | 43 | 0,78 | 1,56 | 57 | 0,86 | 1,73 | 29 | 0,81 | 1,61 | 80 | 0,91 | 1,82 | 91 | 0,84 | 1,69 |
| 45 | 1,00 | 2,64 | 0 | 0,00 | | 104 | 1,00 | 2,71 | 81 | 1,00 | 2,68 | 55 | 1,00 | 3,75 | 66 | 1,00 | 2,95 | 36 | 1,00 | 2,56 | 88 | 1,00 | 2,22 | 108 | 1,00 | 3,26 |

| 23 | Proportions | FI | 24 | Proportions | FI | 25 | Proportions | FI | 26 | Proportions | FI | 27 | Proportions | FI | 28 | Proportions | FI | 29 | Proportions | FI | 30 | Proportions | FI |
|---|---|---|---|---|---|---|---|---|---|---|---|---|---|---|---|---|---|---|---|---|---|---|---|
| 31 | 0,14 | 1,42 | 12 | 0,10 | 0,98 | 9 | 0,09 | 0,87 | 22 | 0,10 | 1,00 | 1 | 0,03 | 0,34 | 0 | 0,00 | 0,00 | 1 | 0,02 | 0,18 | 1 | 0,01 | 0,08 |
| 0 | 0,00 | 0,00 | 0 | 0,00 | 0,00 | 0 | 0,00 | 0,00 | 1 | 0,00 | 0,00 | 3 | 0,10 | 0,10 | 0 | 0,00 | 0,00 | 2 | 0,04 | 0,04 | 4 | 0,03 | 0,03 |
| 188 | 0,86 | 1,72 | 110 | 0,90 | 1,80 | 95 | 0,91 | 1,83 | 196 | 0,89 | 1,79 | 25 | 0,86 | 1,72 | 5 | 1,00 | 2,00 | 52 | 0,95 | 1,89 | 117 | 0,96 | 1,92 |
| 219 | 1,00 | 3,13 | 122 | 100,00 | 2,79 | 104 | 1,00 | 2,69 | 219 | 1,00 | 2,80 | 29 | 1,00 | 2,17 | 5 | 1,00 | 2,00 | 55 | 1,00 | 2,11 | 122 | 1,00 | 2,03 |

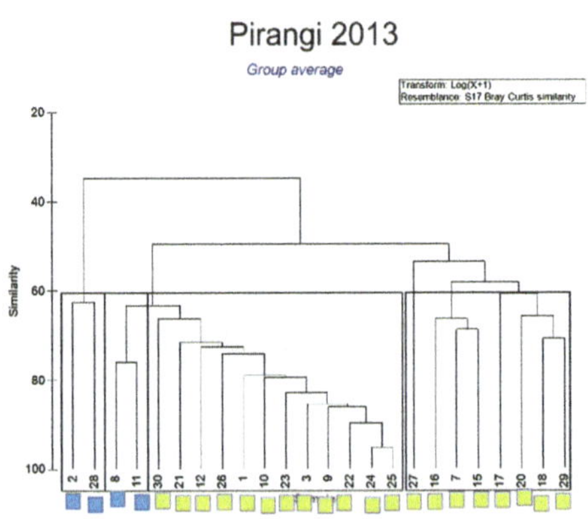


**Fig. 9.15** Cluster analyses and illustrated data from FI for Pirangi 2013

| Pirangi 2014 | 1 | Proportions | FI | 2 | Proportions | FI | 3 | Proportions | FI | 4 | Proportions | FI | 5 | Proportions | FI | 6 | Proportions | FI | 7 | Proportions | FI | 8 | Proportions | FI |
|---|---|---|---|---|---|---|---|---|---|---|---|---|---|---|---|---|---|---|---|---|---|---|---|---|
| SBS | 5 | 0,07 | 0,72 | 29 | 0,35 | 3,49 | 54 | 0,43 | 4,29 | 29 | 0,42 | 4,20 | 29 | 0,33 | 3,33 | 30 | 0,32 | 3,19 | 31 | 0,26 | 2,56 | 36 | 0,35 | 3,50 |
| Oportunists | 0 | 0,00 | 0,00 | 0 | 0,00 | 0,00 | 0 | 0,00 | 0,00 | 4 | 0,06 | 0,06 | 1 | 0,01 | 0,01 | 0 | 0,00 | 0,00 | 0 | 0,00 | 0,00 | 2 | 0,02 | 0,02 |
| small taxa | 64 | 0,93 | 1,86 | 54 | 0,65 | 1,30 | 72 | 0,57 | 1,14 | 36 | 0,52 | 1,04 | 57 | 0,66 | 1,31 | 64 | 0,68 | 1,36 | 90 | 0,74 | 1,49 | 65 | 0,63 | 1,26 |
| Total | 69 | 1,00 | 2,58 | 83 | 1,00 | 4,80 | 126 | 1,00 | 5,43 | 69 | 1,00 | 5,30 | 87 | 1,00 | 4,66 | 94 | 1,00 | 4,55 | 121 | 1,00 | 4,05 | 103 | 1,00 | 4,78 |

| 9 | Proportions | FI | 10 | Proportions | FI | 11 | Proportions | FI | 12 | Proportions | FI | 13 | Proportions | FI | 14 | Proportions | FI | 15 | Proportions | FI | 16 | Proportions | FI | 17 | Proportions | FI |
|---|---|---|---|---|---|---|---|---|---|---|---|---|---|---|---|---|---|---|---|---|---|---|---|---|---|---|
| 24 | 0,32 | 3,20 | 32 | 0,43 | 4,27 | 19 | 0,35 | 3,45 | 18 | 0,38 | 3,83 | 4 | 0,44 | 4,44 | 0 | 0,00 | 0,00 | 23 | 0,36 | 3,59 | 23 | 0,20 | 2,02 | 17 | 0,14 | 1,44 |
| 3 | 0,04 | 0,04 | 0 | 0,00 | 0,00 | 0 | 0,00 | 0,00 | 1 | 0,02 | 0,02 | 0 | 0,00 | 0,00 | 0 | 0,00 | 0,00 | 3 | 0,05 | 0,05 | 0 | 0,00 | 0,00 | 0 | 0,00 | 0,00 |
| 48 | 0,64 | 1,28 | 43 | 0,57 | 1,15 | 36 | 0,48 | 0,96 | 28 | 0,60 | 1,19 | 5 | 0,56 | 1,11 | 9 | 1,00 | 2,00 | 38 | 0,59 | 1,19 | 91 | 0,80 | 1,60 | 101 | 0,86 | 1,71 |
| 75 | 1,00 | 4,52 | 75 | 1,00 | 5,41 | 55 | 0,00 | 4,41 | 47 | 1,00 | 5,04 | 9 | 1,00 | 5,56 | 9 | 1,00 | 2,00 | 64 | 1,00 | 4,83 | 114 | 1,00 | 3,61 | 118 | 1,00 | 3,15 |

| 18 | Proportions | FI | 19 | Proportions | FI | 20 | Proportions | FI | 21 | Proportions | FI | 22 | Proportions | FI | 23 | Proportions | FI | 24 | Proportions | FI | 25 | Proportions | FI |
|---|---|---|---|---|---|---|---|---|---|---|---|---|---|---|---|---|---|---|---|---|---|---|---|
| 14 | 0,12 | 1,24 | 16 | 0,13 | 1,33 | 14 | 0,16 | 1,57 | 12 | 0,23 | 2,26 | 27 | 0,36 | 3,60 | 0 | 0,00 | 0,00 | 33 | 0,35 | 3,47 | 22 | 0,36 | 3,61 |
| 0 | 0,00 | 0,00 | 1 | 0,01 | 0,01 | 1 | 0,01 | 0,01 | 0 | 0,00 | 0,00 | 4 | 0,05 | 0,05 | 0 | 0,00 | 0,00 | 0 | 0,00 | 0,00 | 0 | 0,00 | 0,00 |
| 99 | 0,88 | 1,75 | 103 | 0,86 | 1,72 | 74 | 0,83 | 1,66 | 41 | 0,77 | 1,55 | 44 | 0,59 | 1,17 | 30 | 1,00 | 2,00 | 62 | 0,65 | 1,31 | 39 | 0,64 | 1,28 |
| 113 | 1,00 | 2,99 | 120 | 1,00 | 3,06 | 89 | 1,00 | 3,25 | 53 | 1,00 | 3,81 | 75 | 1,00 | 4,83 | 30 | 1,00 | 2,00 | 95 | 1,00 | 4,78 | 61 | 1,00 | 4,89 |

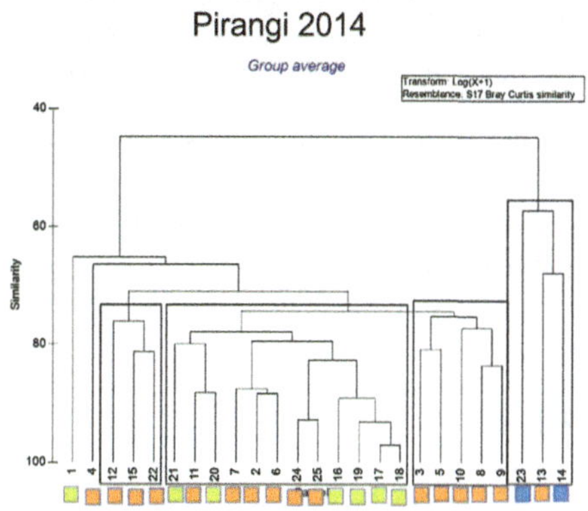


**Fig. 9.16** Cluster analyses and illustrated data from FI for Pirangi 2014

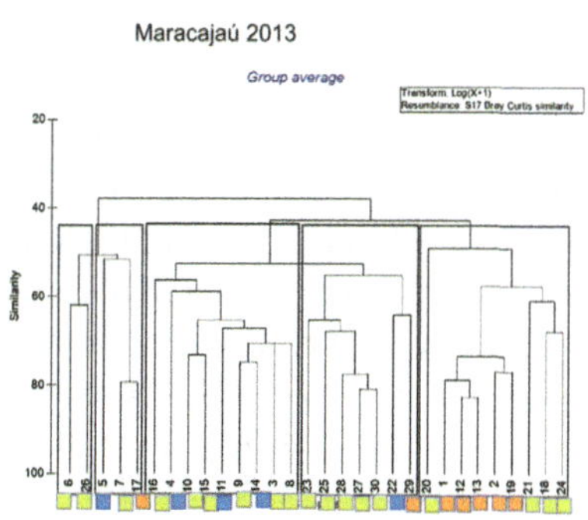


| Maracajaú 2013 | 1 | Proportions | FI | 2 | Proportions | FI | 3 | Proportions | FI | 4 | Proportions | FI | 5 | Proportions | FI | 6 | Proportions | FI | 7 | Proportions | FI | 8 | Proportions | FI | 9 | Proportions | FI |
|---|---|---|---|---|---|---|---|---|---|---|---|---|---|---|---|---|---|---|---|---|---|---|---|---|---|---|---|
| SBS | 48 | 0,53 | 5,33 | 47 | 0,55 | 5,53 | 1 | 0,02 | 0,15 | 0 | 0,00 | 0,00 | 0 | 0,00 | 0,00 | 3 | 0,23 | 2,31 | 3 | 0,18 | 1,76 | 2 | 0,04 | 0,35 | 5 | 0,06 | 0,63 |
| Oportunists | 2 | 0,02 | 0,02 | 0 | 0,00 | 0,00 | 1 | 0,02 | 0,02 | 1 | 0,03 | 0,03 | 0 | 0,00 | 0,00 | 1 | 0,08 | 0,08 | 0 | 0,00 | 0,00 | 2 | 0,04 | 0,04 | 2 | 0,03 | 0,03 |
| small taxa | 40 | 0,44 | 0,89 | 38 | 0,45 | 0,89 | 63 | 0,97 | 1,94 | 36 | 0,97 | 1,95 | 9 | 1,00 | 2,00 | 9 | 0,69 | 1,38 | 14 | 0,82 | 1,65 | 53 | 0,93 | 1,86 | 72 | 0,91 | 1,82 |
| Total | 90 | 100,00 | 6,24 | 85 | 1,00 | 6,42 | 65 | 1,00 | 2,11 | 37 | 1,00 | 1,97 | 9 | 100,00 | 2,00 | 13 | 1,00 | 3,77 | 17 | 1,00 | 3,41 | 57 | 1,00 | 2,25 | 79 | 1,00 | 2,48 |

| 10 | Proportions | FI | 11 | Proportions | FI | 12 | Proportions | FI | 13 | Proportions | FI | 14 | Proportions | FI | 15 | Proportions | FI | 16 | Proportions | FI | 17 | Proportions | FI | 18 | Proportions | FI |
|---|---|---|---|---|---|---|---|---|---|---|---|---|---|---|---|---|---|---|---|---|---|---|---|---|---|---|
| 12 | 0,15 | 1,52 | 0 | 0,00 | 0,00 | 96 | 0,60 | 6,04 | 58 | 0,57 | 5,69 | 0 | 0,00 | 0,00 | 7 | 0,10 | 1,01 | 2 | 0,08 | 0,83 | 6 | 0,35 | 3,53 | 6 | 0,06 | 0,61 |
| 2 | 0,03 | 0,03 | 2 | 0,04 | 0,04 | 1 | 0,01 | 0,01 | 1 | 0,01 | 0,01 | 1 | 0,01 | 0,01 | 1 | 0,01 | 0,01 | 0 | 0,00 | 0,00 | 0 | 0,00 | 0,00 | 3 | 0,03 | 0,03 |
| 65 | 0,82 | 1,65 | 44 | 0,59 | 1,17 | 62 | 0,39 | 0,78 | 43 | 0,42 | 0,84 | 68 | 0,99 | 1,97 | 61 | 0,88 | 1,77 | 22 | 0,92 | 1,83 | 11 | 0,65 | 1,29 | 89 | 0,91 | 1,82 |
| 79 | 1,00 | 3,19 | 0 | 0,62 | 1,21 | 159 | 1,00 | 6,82 | 102 | 1,00 | 6,54 | 69 | 1,00 | 1,99 | 69 | 1,00 | 2,80 | 24 | 1,00 | 2,67 | 17 | 1,00 | 4,82 | 98 | 1,00 | 2,46 |

| 19 | Proportions | FI | 20 | Proportions | FI | 21 | Proportions | FI | 22 | Proportions | FI | 23 | Proportions | FI | 24 | Proportions | FI | 25 | Proportions | FI | 26 | Proportions | FI |
|---|---|---|---|---|---|---|---|---|---|---|---|---|---|---|---|---|---|---|---|---|---|---|---|
| 47 | 0,40 | 3,98 | 7 | 0,24 | 2,41 | 16 | 0,11 | 1,12 | 0 | 0,00 | 0,00 | 7 | 0,04 | 0,37 | 5 | 0,04 | 0,44 | 4 | 0,04 | 0,42 | 1 | 0,11 | 1,11 |
| 0 | 0,00 | 0,00 | 0 | 0,00 | 0,00 | 0 | 0,00 | 0,00 | 9 | 0,23 | 0,23 | 13 | 0,07 | 0,07 | 3 | 0,03 | 0,03 | 10 | 0,11 | 0,11 | 1 | 0,11 | 0,11 |
| 71 | 0,60 | 1,20 | 22 | 0,76 | 1,52 | 127 | 0,89 | 1,78 | 30 | 0,77 | 1,54 | 168 | 0,89 | 1,79 | 105 | 0,93 | 1,86 | 81 | 0,85 | 1,71 | 7 | 0,78 | 1,56 |
| 118 | 1,00 | 0,00 | 29 | 1,00 | 3,93 | 143 | 1,00 | 2,90 | 39 | 1,00 | 1,77 | 188 | 1,00 | 2,23 | 113 | 1,00 | 2,33 | 95 | 1,00 | 2,23 | 9 | 1,00 | 2,78 |

| 27 | Proportions | FI | 28 | Proportions | FI | 29 | Proportions | FI | 30 | Proportions | FI |
|---|---|---|---|---|---|---|---|---|---|---|---|
| 5 | 0,06 | 0,61 | 9 | 0,10 | 1,01 | 1 | 0,30 | 3,03 | 2 | 0,02 | 0,24 |
| 10 | 0,12 | 0,12 | 13 | 0,15 | 0,15 | 3 | 0,09 | 0,09 | 13 | 0,15 | 0,15 |
| 67 | 0,82 | 1,63 | 67 | 0,75 | 1,51 | 29 | 0,88 | 1,76 | 69 | 0,82 | 1,64 |
| 82 | 1,00 | 2,37 | 89 | 1,00 | 2,66 | 33 | 1,00 | 4,88 | 84 | 1,00 | 2,04 |

**Fig. 9.17** Cluster analyses and illustrate data from FI for Maracajau 2013

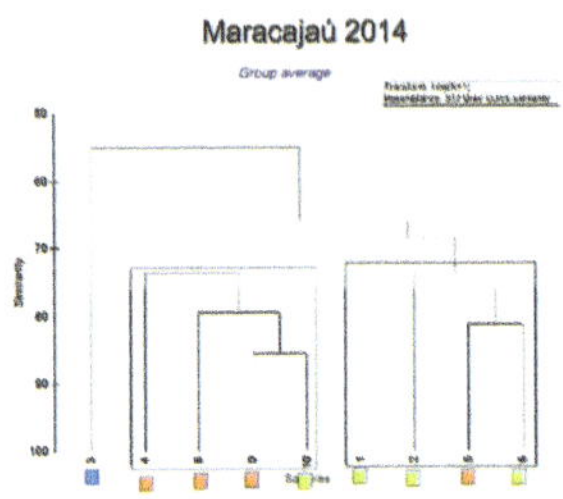


| Maracajaú 2014 | 1 | Proportions | FI | 2 | Proportions | FI | 3 | Proportions | FI | 4 | Proportions | FI | 5 | Proportions | FI | 6 | Proportions | FI | 8 | Proportions | FI | 9 | Proportions | FI | 10 | Proportions | FI |
|---|---|---|---|---|---|---|---|---|---|---|---|---|---|---|---|---|---|---|---|---|---|---|---|---|---|---|---|
| SBS | 13 | 0,13 | 1,26 | 10 | 0,11 | 1,14 | 0 | 0,00 | 0,00 | 26 | 0,41 | 4,13 | 30 | 0,32 | 3,23 | 23 | 0,22 | 2,17 | 34 | 0,35 | 3,51 | 50 | 0,45 | 4,46 | 47 | 0,46 | 4,56 |
| Oportunists | 2 | 0,02 | 0,02 | 10 | 0,11 | 0,11 | 0 | 0,00 | 0,00 | 0 | 0,00 | 0,00 | 1 | 0,01 | 0,01 | 1 | 0,01 | 0,01 | 0 | 0,00 | 0,00 | 0 | 0,00 | 0,00 | 0 | 0,00 | 0,00 |
| small taxa | 88 | 0,85 | 1,71 | 68 | 0,77 | 1,55 | 56 | 1,00 | 2,00 | 37 | 0,59 | 1,17 | 62 | 0,67 | 1,33 | 82 | 0,77 | 1,55 | 63 | 0,65 | 1,30 | 62 | 0,55 | 1,11 | 56 | 0,54 | 1,09 |
| Total | 103 | 1,00 | 2,99 | 88 | 1,00 | 2,80 | 56 | 1,00 | 2,00 | 63 | 1,00 | 5,30 | 93 | 1,00 | 4,57 | 106 | 1,00 | 3,73 | 97 | 1,00 | 4,80 | 112 | 1,00 | 5,57 | 103 | 1,00 | 5,65 |

**Fig. 9.18** Cluster analyses and illustrate data from FI for Maracajau 2014

a tendency of decrease in diversity and evenness with higher dominance of less species today in the eight (8) analyzed cores.

## *FORAM Index (FI)*

FI value higher than 4 indicates an environment "conductive to reef growth and it is represented by peach color. FI between 2 and 4: there must be some symbiont-bearing taxa, and for FI higher than 4, symbiont-bearing taxa must make up at least 25% of the assemblage. Please see illustrative legend for Figs. 9.15, 9.16, 9.17 and 9.18.

The above-mentioned Cluster analysis and data from FI from Pirangi 2013 is shown in Fig. 9.14. Group I: (2 and 28) and Group II: (8 and 11) composed by stations with less number of individuals, no SBS and FI lower than 2 suggests stressed conditions for reef growth. These two groups are in blue. Group III: (30, 21, 12, 26, 1, 10, 23, 3, 9, 22, 24, 25) with coastal stations, medium diversity, greater proportion of SBS; and Group IV: (27, 16, 7, 15, 17, 20, 18, 29) composed by deeper stations, higher number of species are represented by yellow. It is worth to note that this group contains stations (17 and 18), where *Amphisorus hemprichii* has higher proportions. Both groups have FI between 2 and 4 where marginal environment for reef growth is unsuitable for recovery after stress events (Fig. 9.15). Our data showed that *Quinqueloculina lamarckiana* is dominant in blue and *Amphistegina gibbosa* in yellow.

In Pirangi 2014, Group I (12, 15, 22) and Group III: (3, 5, 10, 8, 9) with FI higher than 4 is indicative of an environment conductive to reef growth in orange (Fig. 9.16). Group II: (21, 11, 20, 7, 2, 6, 24, 25, 16, 19, 17, 18); with FI higher than 2 (indicative of an environment conductive to reef growth and marginal environment for reef growth that is unsuitable for recovery after stress events) in orange and yellow (Fig. 9.15). Group IV is composed by (13, 14, 23) a mixture of FI higher than 4, that is indicative of an environment conducive to reef growth, and lower than 2 suggesting stressed conditions for reef growth. Stations 1 and 4 stand by themselves without forming any groups. We have observed higher FI in 2014 than 2013 and dominance of *Quinqueloculina lamarckiana* in blue sites, *Homotrema rubra* in yellow, and presence of *Amphistegina gibbosa* and *Borelis schumberger* in Orange. We could also observe that FI is lower in a touristic area visited by boats and where people swim, dive and walk on the reefs.

In Maracajau 2013, Groups I (6, 26) and II (5, 7, 17), with FI between 2 and 4 and where marginal environment for reef growth is unsuitable for recovery after stress events, and lower than 2 are stressed conditions for reef growth. Both groups with only *Amphisorus hemprichii* in yellow and blue. Group III: (16, 4, 10, 15, 11, 9, 14, 3, 8) with FI between 2 and 4 Marginal environment for reef growth that is unsuitable for recovery after stress events and lower than 2 is stressed conditions for reef growth, with few specimens. Group IV: (23, 25, 28, 27, 30, 22, 29) with FI between 2 and 4 where marginal environment for reef growth is unsuitable for recovery after stress, however higher diverse stations with *Amphisorus hemprichii* and events in yellow. Group V: (4, 18, 9, 10, 20, 1, 12, 13, 2, 19, 21, 18, 24) with FI higher than 4, and between 2 and 4 indicative of both an environment conductive to reef growth, and a marginal environment for reef growth that is unsuitable for recovery after stress events where *Amphistegina* dominates (Fig. 9.17).

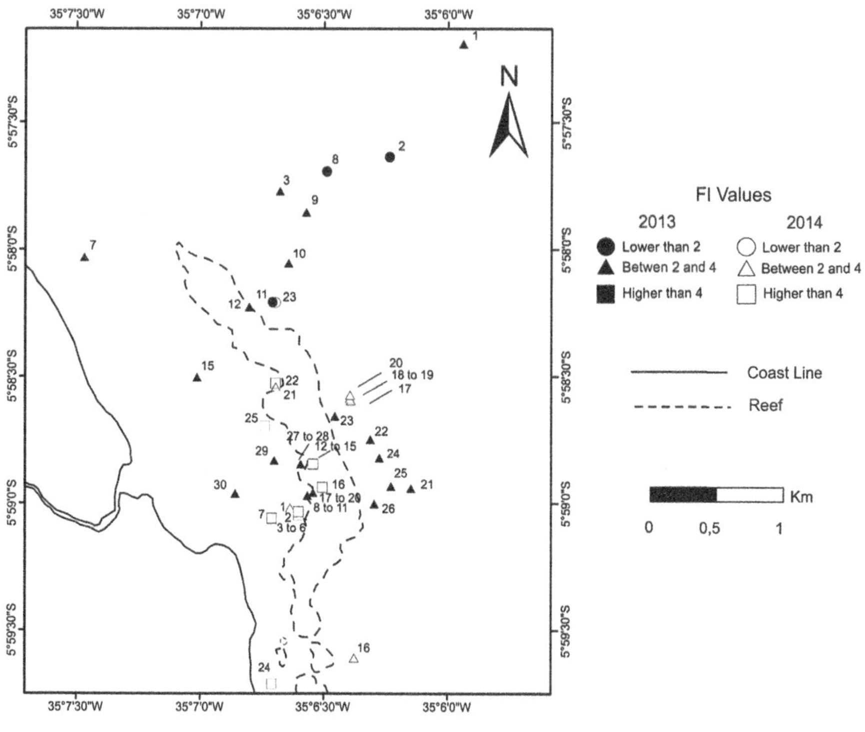


**Fig. 9.19** FI values for Pirangi reef area

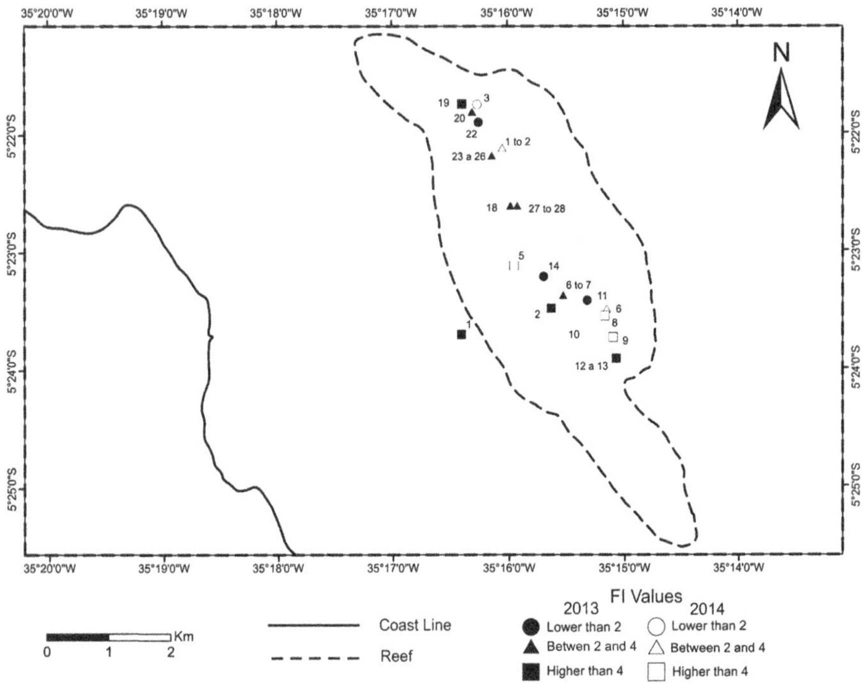


**Fig. 9.20** FI values for Maracajau reef area

In Maracajau 2014, Group I: (1, 2, 5, 6) with FI between 2 and 4 where marginal environment for reef growth is unsuitable for recovery after stress events, diverse stations and the presence of *Amphisorus hemprichii* in yellow, and Group II: (4, 8, 9, 10) mostly higher FI than 4, indicative of an environment conductive to reef growth in peach. They are less diverse stations with dominance of *Amphistegina*. Station #3 FI is lower than 2, with stressed conditions for reef growth, and stands by itself because there is just one species (*Quinqueloculina lamarckiana*) living in this area (Fig. 9.18).

Representation of FI data on illustrative Fig. 9.19 show that in Pirangi samples were collected rather out of the reef, since most of Pirangi reef area emerge and is more exposed to air, so it was not sampled for this study. Higher FI values were found in 2014 mostly in the southern part in the back-reef areas less exposed to wave energy.

Maracajau FI data in Fig. 9.20 shows also higher values for the southern part in both years. Marcajau reef area is submersed and only one station was collected in the back reef.

## Discussion

Local variables such as high hydrodynamics and other reworking processes might influence the faunal occurrence of Forams in this portion of the shelf and therefore alter the interpretation of the FI index. This index should be applied with caution because the high occurrence of symbiont-bearing Forams could lead to the interpretation of high-water quality, which is not necessarily true as already shown by Barbosa et al. (2009) in a comparison with coral reef coverage in different sites of the Brazilian coastal area. In the same manner, low FI values correlates with silt in the sediment (not suitable for symbiont-bearing foraminifera), and not with low water quality for coral survival in the Abrolhos area, as these sites have the highest coral cover (30–32%) considered as some of the best reefs in the world (Barbosa et al. 2006; Ferreira and Maida 2006). Oliveira-Silva et al. (2012) showed that foraminiferal assemblage appears to be responding most strongly to sediment texture, algae and coral cover, which influences the proportion of functional groups), where relative frequencies of hard coral and algal cover from Abrolhos and Corumbau sites demonstrate that coral communities are well developed in these regions. However, FI numbers show water quality might be declining. Silva et al. (op. cit) also revealed that FI can be strongly influenced by sediment texture and should be interpreted accordingly.

In our study, dominance of smaller taxa, including stress-tolerant species, and minimal representation of symbiont-bearing taxa, indicate unsuitable conditions for these important calcifiers in both Pirangi and in Maracajaú. Our work also demonstrates that foraminiferal fauna correlates well with coarse and sand fractions and depth was the least influencing variable. We are also cautious when applying a bio indicator developed in one region to a new region, because there are regional differences in the adaptability of coral communities, and sampling methodology.

Based on Forams assemblages in Brazilian reef environments, FI provided reliable proxy for water quality and reef health. They were tested to determine if FORAM Index (FI) indicates that water quality supports benthic communities of calcifying symbionts. Barbosa et al. (2012) found the lowest values in the harbor area where corals were sparse. The same occurs in Pirangi, Maracajau where boats dock onto, and tourists "step on" coral reefs. In addition, the absence of living *Amphistegina* specimens and its taphonomy study showed that coral communities might be at risk.

As observed by Araújo and Machado (2008), the low diversity and evenness index found at Abrolhos explained the dominance of *Quinqueloculina, Amphistegina* and *Archaias* in shallow, low-energy stations and this dominance increased FI. FI results are different from those found by Schueth and Frank (2008) in Australia at similar protected sites, suggesting that in Brazil depositional energy, besides depth, seems to control the distribution of large symbiont-bearing foraminifers. Our results confirm their findings since our higher diversity is from smaller taxa, and opportunistic, and not from symbiont bearings species. In addition, our low evenness values may be related to the transport and selection of tests by bottom currents, which prevent Forams settlement. In general, the FI reached high values in sandy sediment, whereas low values were associated with muddy sediments, as observed by Barbosa et al. (2009). Sanches et al. (1995) concluded that *Archaias* was the most abundant and frequent genus (24% in 1995 against 9% in Barbosa et al. 2009) of the foraminifer community at Abrolhos, occurring at all stations studied. The change in the community has been observed, wherein *Quinqueloculina* has become the dominant taxa (25% in Barbosa et al. 2009) against 12% of the assemblage in 1995). Cockey et al. (1996) also observed a change from symbiont bearing to heterotrophic foraminifers like *Quinqueloculina* when studying Florida reef systems. The genus *Amphistegina* occurred mostly in ($10 \leq x \leq 15$ m) and ($15 \leq x \leq 25$ m) domains, which is in agreement with Baker et al. (2009) that *Amphistegina gibbosa* was always dominant at depths of 10–30 m at Florida reef. Our study also shows dominance by *Quinqueloculina, Amphistegina* and *Archaias* in shallow reef areas of Pirangi and Maracajaú. Pirangi's samples were a mixture of recent and relict tests, the latter were not only brownish and yellowish, but also eroded and broken, suggesting they may be relict sediment, as observed by Moraes and Machado (2003) for Forams on the carbonate shelf of Bahia state and by Machado and Souza (1994) at Rocas atoll (Rio Grande do Norte State).

Cluster analysis reveal clear patterns of grouping sites among stations, and could be linked to ecological indices differences. The samples were not grouped by depth, suggesting that regional group prevalence driven mainly by wave energy and a degree of exchange with open marine waters (both not measured) influenced the foraminifer distribution via water and sediment quality, as seen by Narayan and Pandolfi (2010) in a subtropical estuarine environment. On the other hand coarse fraction are the best-correlated variable. Cluster identified clear separation within the data where sites are grouped in relation to its indication of conduciveness to reef growth, or not, after stress events.

PCA analysis show two first principal components represent up to 67.7% of the total for Pirangi 2013 and 76.5% for Maracajau in 2013. Highest positive values for PC1 in both reef areas are explained by percentage of sand availability, and in Maracajau variability of samples is also explained by percentage of $CaCO_3$. Highest negative values of PC1 are silt and clay for both reef areas also contributed to the occurrence of other heterotrophic (*Bolivina, Elphidium, Nonion*) and stress-tolerant foraminifera (*Cibicides, Discorbis, Miliolinella, Pyrgo, Quinqueloculina, Rosalina,* and *Triloculina*). In Maracajau, total organic matter also contributed to the occurrence of the heterotrophic and stress-tolerant foraminifera genera.

The low frequency of stained foraminifers in both reef areas might be related to the same factors listed by Bicchi et al. (2002), such as: (1) empty tests should be common in reef areas; (2) most reef-dwelling foraminifers live attached to reefs, stalks and macroalgae, and only empty tests should be found free in the sediment after a reproductive event. Thus the sediment assemblage, although modified by postmortem processes, is an averaged mosaic of spatially and possibly temporally varying communities, as shown by Wilson and Ramsook (2007) in the West Indies; our results also agree with that, especially in Pirangi.

The mean FI values at Pirangi ranged from less than 2 to 4 at most stations, indicating that Pirangi is a marginal environment for reef growth and unsuitable for recovery after stress events according to the index interpretation proposed by Hallock et al. (2003). Our results agree with Hallock (op. cit.), especially in areas with touristic activities; however, (Barbosa et al. 2009) does not agree and despite the low FI values at some sites, this does not reflect the real condition of Abrolhos in the eastern Parcel area which presents remarkable coral coverage (around 32%) and high reef fish diversity. The present study can contribute to the management plan of the Pirangi and Maracajaú National Marine Parks by applying knowledge of its foraminifer assemblage to the diagnosis of the environmental health of the region, and by providing a first insight to the pattern in which organic matter is distributed in the sediment across depths.

By associating the foraminifer composition with the geochemical data, we can conclude that coarse and sand fraction are the controlling parameters. It is also clear that depositional energy plays an important role in transportation and deposition of sediments and foraminifera. The dominance of *Quinqueloculina* (heterotrophic) shows changes that have yet to be identified are occurring in the area and must be taken into consideration in further studies. According to the FI results, the environmental health evaluation shows that the water quality of the Pirangi is not suitable for coral reef growth, and Maracajau has sites suitable for coral reef growth even if it has individual sites where coral could not survive after a stress event. Heavily reworked individuals were avoided in order not to negatively influence the index, as the majority of the foraminifer assemblages in this area suggest a relict character of the sediments. Long-term assessments are in need in order to improve our knowledge regarding the distribution and ecological importance of Brazilian reef-dwelling foraminifers, as well as to extend the application of the FI to large-scale monitoring of this and other reef ecosystems in the Southwestern Atlantic.

A Taxonomic study performed in the University of California Museum of Paleontology (UCMP) from August 24th to September 8th, 2015 was important due to the necessity of looking at true reef foraminiferal species and be sure we had the right species of all genera, and some studies were done at UCMP to investigate this. The *Amphistegina* study is illustrated in Fig. 9.21.

The *Amphistegina* study revealed that specimens from Coral reef in Rio Grande do Norte State is indeed *Amphistegina gibbosa*. We could also verify that *A. gibbosa* is common in Bahamas, Philippines and Tonga islands and similar from RN reef areas being the subject of our study on Fig. 9.22.

Larger foraminifera genera and species from Marshall Islands, Philippines, Bahamas, Australia, Tonga Islands, Samoa, Hawaii, Ifaluk (a coral atoll in the Pacific Ocean), East Indies (Figs. 9.23 and 9.24).

A detailed study was done on Ifaluk region, Micronesia. There are similarities with the microfauna from RN reefs: *A. gibbosa, Peneroplis carinatus, and Bolivina* sp. were found however, and some species of *Calcarina* and *Sorites* are absent in our RN reefs (Fig. 9.25).

Reef areas in Honolulu (Hawaii), Java (Indonesia), New Guinea (Philippines) are illustrated in Fig. 9.26.

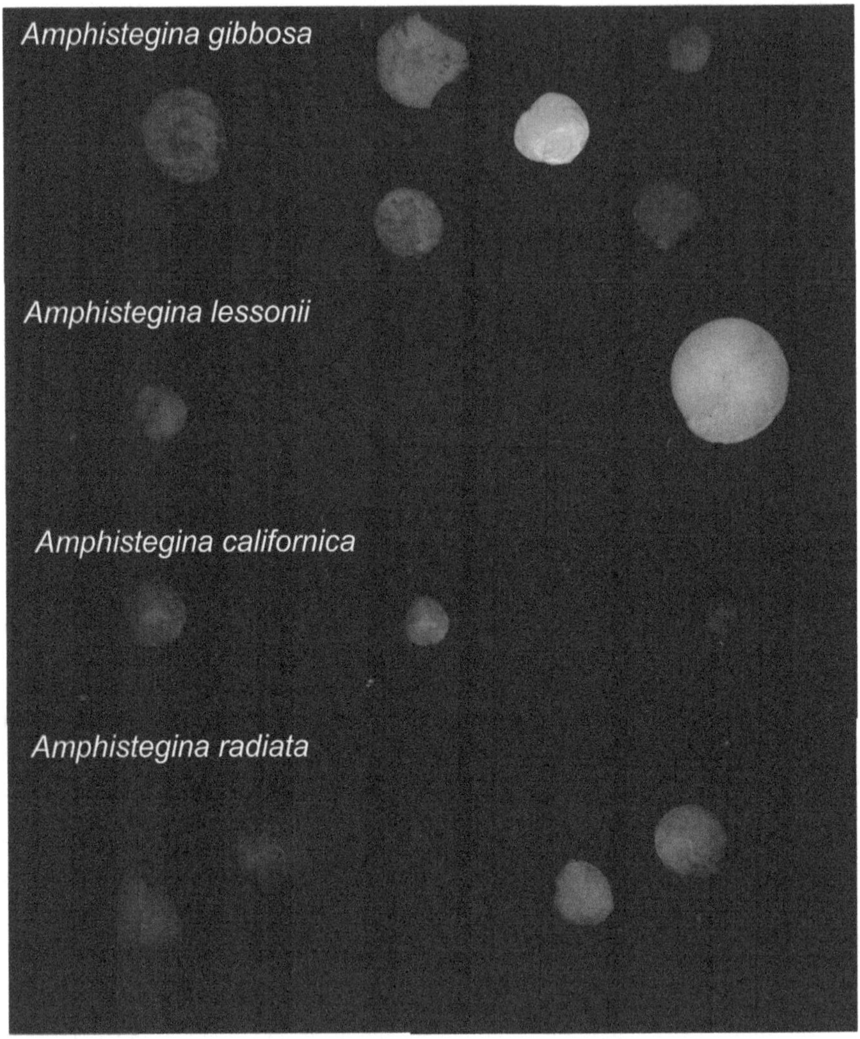


**Fig. 9.21** Different species of *Amphistegina*

**Fig. 9.22** *Amphistegina* and other genera from different localities

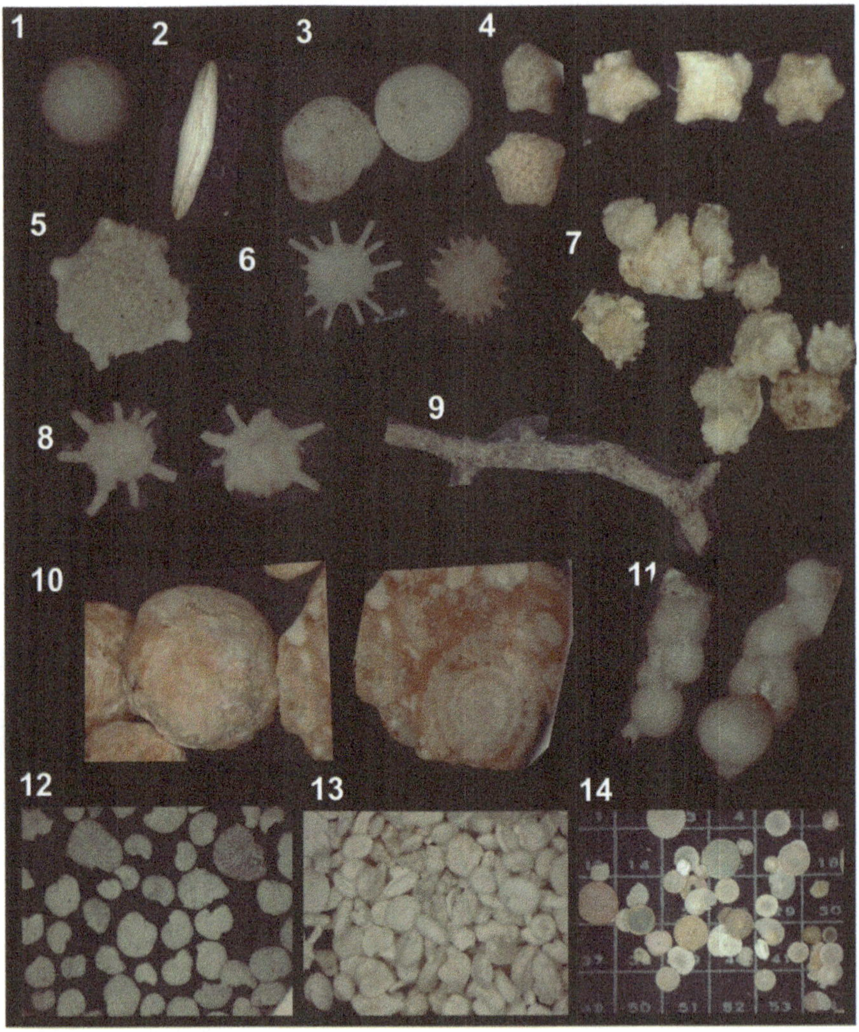


**Fig. 9.23** Larger Forams genera from all over the world. 1. *Acervulinidae,* Marshall Islands 2. *Alveolinella boscii,* Philippines 3. *Archaias angulatus,* Bahamas 4. *Baculagypsina sphaerulatus,* Australia and Tonga Islands 5. *Calcarina hispida,* Philippines 6. *Calcarina mayoni,* Samoa 7. *Calcarina spengleri,* Tonga Islands 8. *Calcarinidae,* Marshall Islands 9. *Dendrophyra attenuate,* Philippines 10. *Camerina*, Tonga Islands 11. *Carpenteria proteiformis,* Philippines 12. Slide from Bahamas 13. Slide from Australia 14. Slide from Hawaii

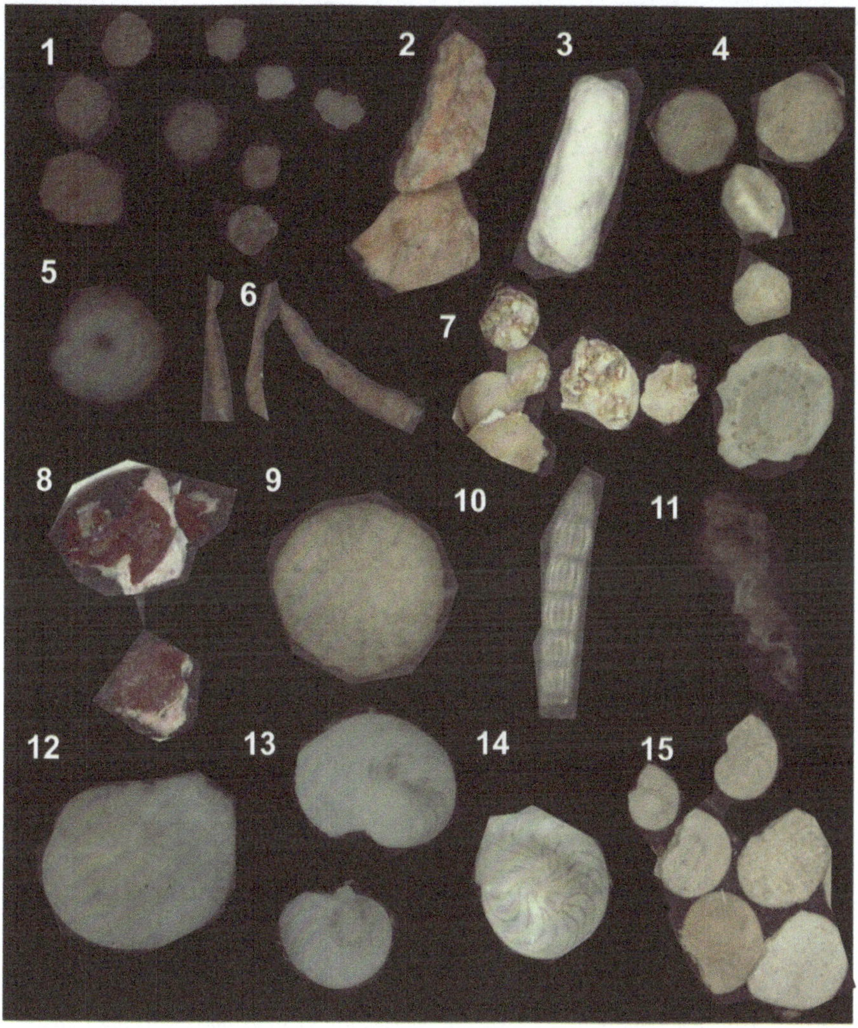


**Fig. 9.24** Larger Forams genera from all over the world. 1. *Cymbalopora bradyi,* Samoa 2. *Discocyclina*, Tonga Islands 3. *Eleuthera*, Bahamas 4. *Gypsina vermicularis*, Tonga Islands 5. *Fishcherinidae*, Marshall islands 6. *Haliphysema catenulate*, Philippines 7. *Heterostegina depressa*, Tonga Islands, 8. *Homotrematidae*, Marshall islands 9. *Marginopora*, Ifaluk 10. *Nodosaria vertebrata*, pacifico, 11. *Nubeculina divaricata Variation advena*, Samoa 12. *Numulitidae*, Marshall islands 13. *Operculina bartsehi*, Philippines 14. *Operculina cumingii*, East Indies 15. *Operculina philippensis*, Philippines

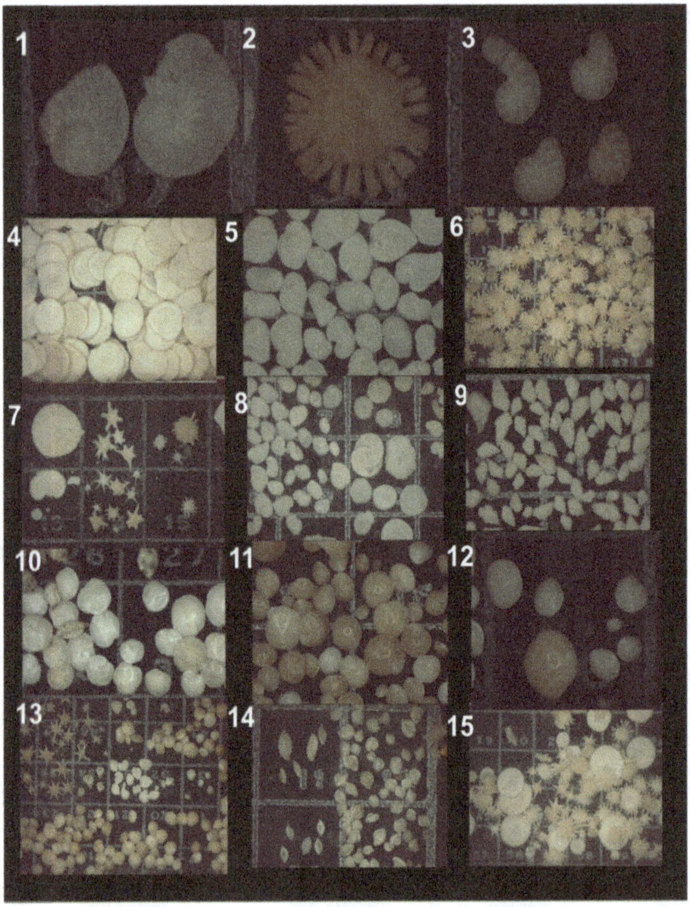


**Fig. 9.25** Larger foraminifera genera from IFaluk island. 1. *Heterostegina depressa*, 2. *Calcarina* sp., 3. *Peneroplis pertussis* (1 specimen) and *P. carinatus* (3 specimens), 4. Soritidae, 5. *Peneroplis planatus, P. pertussis*, 6. *Calcarina* sp., 7. Soritidae and *Calcarina* sp., 8. *Peneroplis planatus, Sorites* sp. 9. *Bulimina* sp. 10. 11. 12. *Amphistegina gibbosa,* 13. *Calcarina* sp., *Amphistegina gibbosa, Bulimina, Peneroplis*. 14. *Quinqueloculina* sp. 15. *Calcarina* sp. and *Sorites* sp.

Reef areas in Philippines, East Indies, Bahamas, Australia, and Marshall islands are illustrated in Fig. 9.27.

## Recommendation

Suggestion for delimitation of permanent conservation areas for the management of the Pirangi Marine National Park (RN, Brazil) to increase biodiversity. This chapter proved our hypothesis right, verifying that foraminifera can be used as "health bio-

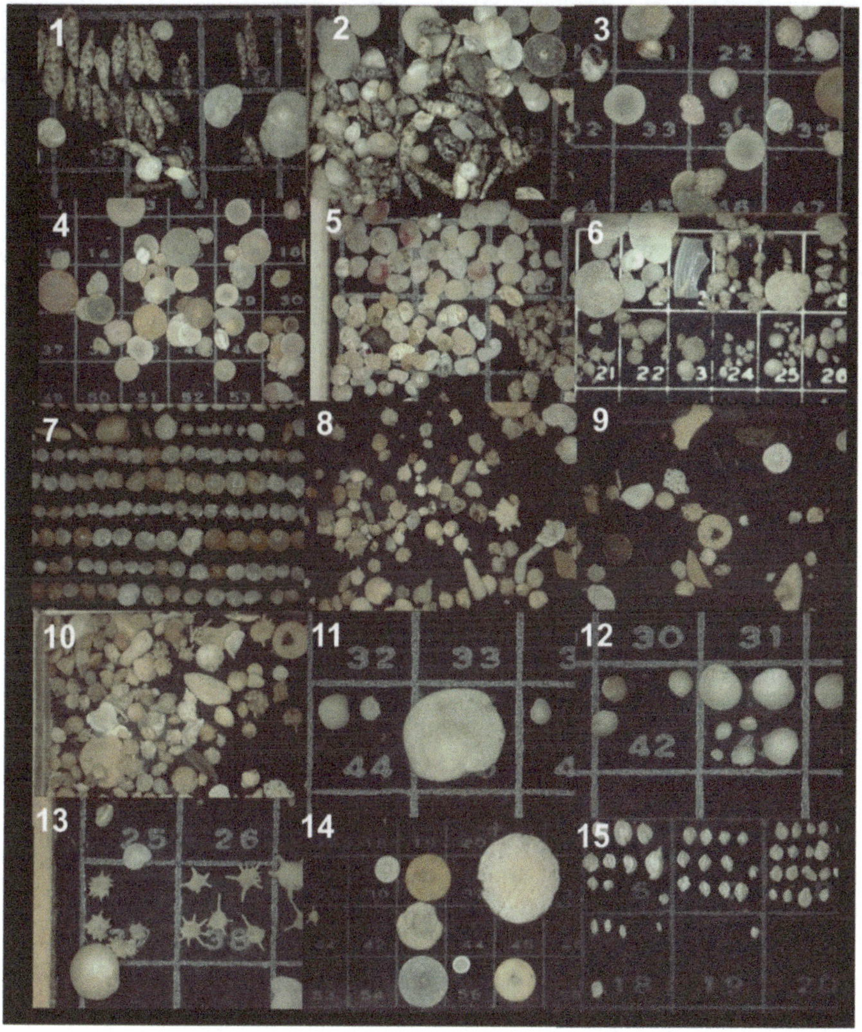

**Fig. 9.26** Foraminifera genera from Reef areas in Honolulu (Hawaii), Java, (Indonesia), New Guinea and Philippines. 1. *Lituola* (agglutinated) and *Operculina* sp. (Hawaii), 2. Soritidae, Lituola and *Operculina* sp. (Hawaii), 3. Soritidae and *Amphistegina gibbosa* (Hawaii), 4. Soritidae (Hawaii), 5. *Peneroplis* and *Lituola* (Hawaii), 6. *Operculina* sp. (Indonesia), 7. Indonesia (mixed fauna), 8, 9, 10 New Guinea (mixed fauna), 11 *Heterostegina depressa*, 12. *Amphistegina gibbosa*, 13 *Calcarina* sp. and *Amphistegina* (Philippines), 14 *Soritidae*, 15 *Quinqueloculina* (Philippines)

indicators of reef areas". Our objective was to evaluate the local anthropic stresses in the ecosystems of little known on the Brazilian coast, the reefs of Maracajaú and Pirangi of RN—through the study of the structure of the associations of benthic foraminifera present on the surface of sediments in the vicinity of corals.

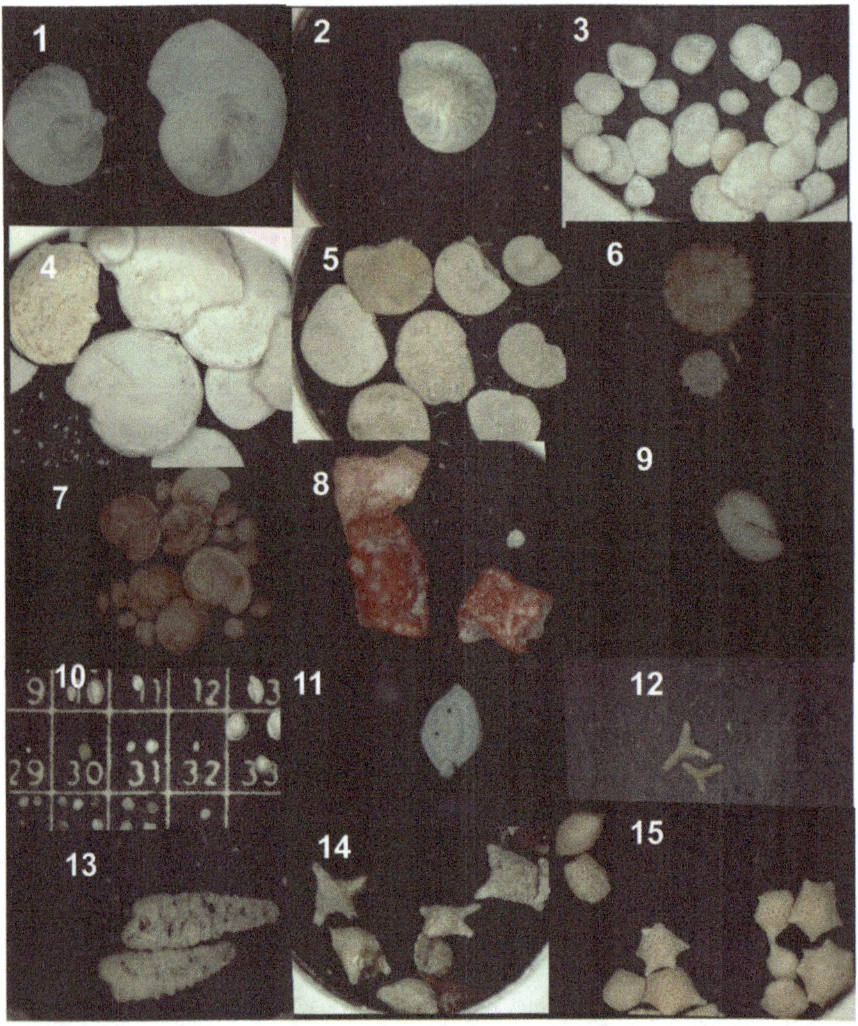


**Fig. 9.27** 1. *Operculina bartsehi* (Philippines), 2. *Operculina cumingii* (East Indies), 3. 4. *Orbiculina adunca* (Bahamas), 5. *Operculina philippensis* (Philippines), 6. *Planorbulina larvata* (Philippines), 7. *Orbiculina achemea* (Bahamas), 8. *Polytrema mimeaceum* (Murray island, Australia), 9. 10. *Quinqueloculina costata* (Bahamas), 11. *Spiroloculina* (Marshall Islands), 12. *Saccorkiza ramose* (Philippines), 13. *Textularia goesii* Philippines, 14. *Siderolites tetraedra* (Philippines). 15. *Tinoporus baculatus* (Murray island, Australia)

Our results show that among the two reef areas studied the Pirangi area seems to present the worst indices of diversity and stability of the environment, with dominance of opportunistic species, and little or absence of species of symbiotic foraminifera with algae. The absence of these species denotes a decline in fauna, as they are indicators of environments of excellent ecological health. Specifically, close to

**Fig. 9.28** Sampling stations analyzed for foraminifera in Pirangi 2013 and 2014. Area in red: suggestion for delimitation of Permanent Preservation Areas

the commercial exploration area, the levels of diversity and equitability tend to drop greatly revealing that the trampling of algae and the benthic community as a whole is in sharp decline, and that decline tends to spread in areas situated nearby.

Our data show that the delimitation of an area for permanent conservation is of utmost importance in the remediation and preservation of the marine area of the Pirangi reefs, as a source of propagules. The suggested area covers the areas that are currently the most productive in our diversity and stability of the environment. Figure 9.28 shows the sediment and water sampling stations in the Pirangi area and the suggested delimitation for "isolation for preservation" area.

# Chapter 10
# Microfossils as Saviors?: How to Prepare for Climate Changes… Using Microfossils?

**Abstract** Pure & Simple: We study ocean sciences, to understand sea-land processes, communicate alterations caused by global climate (warming & cooling) changes and other environmental problems, collaborate with colleagues globally, and accurately answer questions about challenges we face in the near future, with recommended solutions! With a local presence, global reach and cost effective approach, EcoLogic Project befor (ELP) offers the best, most efficient oceanographic, coastal and environmental consulting services in the industry. Our services include vulnerability assessment consultancy, planning, communications, and project management. Also data acquisition, surveys and laboratory analysis. Additionally, we offer educational presentations for the public, private entities, government and municipalities. While adhering to critical project deadlines and milestones, we develop strategies to inform leadership with knowledge and data to effectively pinpoint problematic climate challenges and potential loss. EcoLogicProject unravels the mystery of the effects of Climate Change with Environmental Impact Consultancy and Advisory Services, helping communities prepare, adjust, adapt, and minimize effects or remediate loss. Are you wondering how climate change will affect your family, friends, community, and city? For inquires, presentations, outlined or detailed Project Assessment and Cost Analysis, please contact us to address your needs for a customized proposal specific to your project. Please send your questions to us at ELP website.

**Keywords** EcoLogicProject · Global · Ecological authority · Preservation · Climate change · Predictable indicators · Loss · Remediation · Mitigation · Advisory

 
P. P. Beck Eichler, C. P. Barker, *Benthic Foraminiferal Ecology*,
https://doi.org/10.1007/978-3-030-61463-8_10

## Introduction

A global ecological stewardship authority is needed to prepare for Climate Change Affects. The entirety of humanity is failing as it continues toward its own destruction through negligence to integrate gently in cooperation with nature's balance. We need benevolent leadership that is noble of character placing dignity in moral integrity above profit, to educate, promote, unite and build a shared ambition that inspires unity in all cultures for the conservation and preservation and integrative sustainability needed to thrive within our interdependence of the web of life.

Approximately 70% of the Earth's surface is covered by oceans, and life likely emerged from there around 3.5 billion years ago. Oceans act as the planet's thermostat, transferring heat from the tropics to higher latitudes and absorbing a substantial amount of the carbon dioxide that the world's plant life does not convert into oxygen through photosynthesis.

A comprehensive all-inclusive cooperative global social awareness and stewardship action is needed for the enhancement of biodiversity as a whole. Through enlightened leadership, integrative research, study, and action, we can ensure the protection and pristine restoration of all ecosystems needed to sustain viability for our children, all future generations, and ourselves.

Ignoring the laws of nature—of cooperative sustainable interconnectivity, and replacing this stewardship instead with greed has eroded even the most basic inalienable human and environmental rights. Our collective ignorance and indifference of myopic malaise has led us to the edge of what may well be an irreversible "Anthropocene era 6th extinction".

The ultimate objective of 'unite and build' leadership is not for gaining power and money, but to empower each other to achieve something bigger through collaboration and sharing resources to better preserve nature. Competition for, and depletion of, natural resources, in favor of a few people, does not consider the vital importance of our dependence on all life, and predictably will always lead us to failure. When we are united, we can successfully surmount the challenges of biodiversity loss and climate change, and dramatically alter the current trajectory to extinction. With comprehensive planetary planning that result in effective strategies, we will enable continuity for all living beings, which rely on biodiversity, including humanity, to live with freedom, comfort and dignity, without negatively influencing the earth's ecosystems or regenerative ability.

Simple logic shows that pollution and extinction are not parallel lines; eventually they will meet, and that timeline appears to be induced, enhanced and accelerated by human activity. And that predictable end is in rapid approach in this Era. The growing body of facts and evidences demonstrate that widespread ecological collapse and biodiversity loss is therefore eminent. As average temperatures rise now, the frequency of extreme events increase, and the impacts of habitat loss and fragmentation will become even more increasingly apparent as well.

We see problems now with food security, as an example the wildlife market in China and the new coronavirus, Covid 19, which besides instability of ecosystem supply chain poses threats to our old and now outmoded way of life, and of course wildlife and sustainable developments are at risk. It is mandatory that we restore and reconnect our landscapes and watersheds, reduce and eliminate emissions, prevent worldwide species extinctions, sequester carbon, and steward biodiversity to stop the sixth mass extinction event. These problems are accelerating at an unprecedented rate, and kicking off disturbing effects, posing a real threat all life including to human existence.

Humankind is simply one of countless lifeforms that rely heavily on biodiversity and healthy ecosystems to survive and flourish. As species disappear due to habitat destruction, wildlife trade, pollution and climate change, so do the essential parts of a delicate system we depend on for our well-being. When humanity exterminates populations and species of other creatures, it is sawing off the limb on which it is sitting, destroying working parts of our own life-support system. The conservation of endangered species should be elevated to a national and global emergency for governments and institutions, equal to climate disruption to which it is linked (Ehrlich 2015). This refers to extinction of one species impacting other species that rely on it for survival, placing them at a 'domino effect' risk of extinction as part of a destructive chain reaction ending on species termination in different trophic levels affecting the ecological chain.

Researchers propose a number of ways in which we can address the problem, including species and regions where preservation and conservation resources can be best directed, highlighting which factors are most impactful in driving these accelerating extinction rates. This research highlights the fragility of the Earth's support systems and the urgent need to act. The call for the preservation of endangered species to be elevated to a national and global emergency is both warranted and urgent (Clarke 2020).

As we have done with EcoLogicProject.com, our endeavor here is to provide accurate data, with predictable indicators for climate change including but not limited to ocean warming, cooling, and acidification. Our goal at "EcoLogicProject" (ELP) is to add to an enlightened paradigm narrative for humanity's progress toward sustainable 'eco-stewardship'. To provide readable reliable accurate information about environmental factors in changing ecosystems, and to demonstrate how global climate change will affect future sea level rise, and environmental, social and economic loss using microfossils.

In addition, we identify patterns and potential problems that may arise now, or in the near future, and we endeavor to make sound recommendations for people to understand and adapt, to make the necessary changes on a warming planet where warmer oceans will expand in addition to water added by melting glaciers and polar areas based on evidences and facts.

The recommendations for change that we offer here range from relocation, and original reforestation and restoration, to collection of debris, and construction of monitoring plans due to climate change effects already worldwide. Shrinking glaciers, and the shifting of latitudinal animal and vegetative cover because of varied tolerance ranges, are evidences of the fact that it is happening in real time, with predictions now coming to fruition through loss of sea ice, intensified heatwaves and hurricanes, and measurable sea level rise currently around the world.

Temperatures will continue rising for "decades to come" primarily due to human-caused greenhouse gas emissions. Forams are the first consumers and they are the first ones to show if there is a problem locally and globally in the ecological chain, and in the environment as a whole. This is because they show oceanic-source erosional and depositional sites, as well as the deposition of organic matter and other materials coming from land and are statistically accurate and trustable.

The most recent report for the First Order Draft of the IPCC Special Report on Climate Change and Land (SRCCL) published by the Intergovernmental Panel on Climate Change (IPCC), which Professor Eichler is an Expert Reviewer, confirms findings predicting a 1.5 °C increase of temperature above pre-industrial levels around the world. This effect is already bringing more droughts and heat waves, changes in precipitation patterns, stronger more intense and more frequent hurricanes, and an expected sea level rise of between 1 and 4 ft. by the end of this century, all coming from warming and acidification of water and marine sediments evidences. Dr. Eichler recommended, as a part of the IPCC studies from now on, measurement of ocean acidification rate as it poses a problem to all marine life.

Similar past impacts are recorded in marine sediments with evidence unraveled by foraminifera as microfossils, which therefore detect near future local specific problems, giving time for people to understand with evidence to make changes and prepare for what is a "enviro-collapse" bullet-train coming fast in our direction.

Regarding the climate change increase that is coming, we will include in our conversation the 20 Corporations that most pollute the ocean, and that need to be held accountable!

These corporate-influenced changes to the ocean chemistry, currents, and temperatures have profound effects on global weather patterns, and human industrial and consumption activity is adversely impacting the oceans and their complex ecosystems. The result is a number of species that are very sensitive cannot adapt to or tolerate anthropogenic toxic overload inputs, and therefore are becoming extinct.

Pollution from plastic has received much attention in recent years as people cringe at images of tropical beaches covered in garbage. Plastic bags and bottles cause animal deaths, and massive garbage patches are found in all of the world's largest bodies of water. These debris fields consume all the oxygen in previously healthy areas, generating sites where eutrophication is high and includes the mortality of fishes and other benthic communities. This endangering plastic problem, together with the micro plastics entering the ecological chain has helped raise awareness of ocean pollution; however, this is just one of the problematic issues. Threats to our ocean's health include greenhouse gas emissions, and especially excessive carbon dioxide, which we are releasing into the atmosphere. These emissions are related to, but not limited to global warming; causing ocean acidification and the consequence is the dissolution of calcareous organisms like foraminifera, crabs, mollusks and many other species that need a $CaCO_3$ marine environment in equilibrium to be able to form their tests. This dissolution is devastating to specific species and to marine ecosystems in general. In addition to carbon emissions and plastics pollution, oceans are also polluted and contaminated from oil and chemicals and sewage that wash into rivers and wind up in the oceans. This non-point source pollution includes oil leaked from cars, trucks, and runoff from residential and industrial areas, and farms and ranches that use herbicides, pesticides, and fertilizers flowing into the oceans via rainwater runoff into rivers. When burned, fossil fuels emit greenhouse gases, which also cause ocean acidification, a harmful process that also disrupts marine ecosystems and life on 70% of the planet—"the ocean".

A list of the 20 companies that most pollute the oceans regarding this above-mentioned major sources was compiled by the 24/7 Wall St. It includes major producers of single-use plastic containers, fossil fuel companies, meat and dairy producers, and agricultural chemical manufacturers responsible for the largest share of oceanic contamination, on their relative size in industries that are known to pollute heavily, as well as estimates of these companies greenhouse gas emission effects by third parties.

Here we will talk about them briefly, and why they are polluting. 24/7 Wall St. reviewed information about different sources of pollution: plastic pollution; runoff from agricultural pesticides, herbicides, and fertilizers; and greenhouse gas emissions and discuss some of the issues below. For the primary company sources of plastics pollution, 24/7 Wall St. used a 2018 study of plastic garbage collected in 42 countries and catalogued by Greenpeace and the Break Free from Plastics Movement. For the primary sources of greenhouse gases—oil, gas, and coal companies responsible for the most greenhouse-gas emissions—they reviewed data from 2017s Carbon Majors Database from the Carbon Disclosure Project (CDP), a UK-based charity encouraging carbon emissions reporting. The database considered here is from 2015. Also, a 2018 study from the international nonprofit organization GRAIN and the Institute for Agriculture and Trade Policy (IATP), a non-profit focused on environmental sustainability, provided data on the largest meat and dairy companies that release the most greenhouse gas emissions. They used a 2018 report from Agrow, an agribusiness intelligence company, to find the biggest producers of agricultural chemicals based on sales, which is how they determined the largest polluters in the industry. The U.S. National Oceanic and Atmospheric Administration provided background on so-called non-point source pollution, the ocean pollution that comes from runoff from vehicular oil leaks, farms and ranches and so on.

Saudi Aramco: The state-owned Saudi oil and gas giant is the world's most profitable company. It is also the biggest emitter of greenhouse gases in the fossil fuel sector. In addition to being the industry leader in GHG emissions from its own operations, Saudi Aramco is the overall leading supplier of fossil fuels in the global energy sector.

Gazprom: The state-owned gas giant and Russia's most valuable publicly traded company is the second largest source of greenhouse gases, which causes ocean acidification, among global fossil fuel companies based on 2015 data. It ranks third among the top energy companies in terms of greenhouse gas emissions from operations, but the burning of the petroleum liquids and natural gas it produces and sells puts it in second place in overall greenhouse gas emissions, after Saudi Aramco.

National Iranian Oil: ranks second in greenhouse gas emissions from its operations, and third in total emissions linked to its operations and its products, behind Saudi Aramco and Gazprom. Recent U.S. sanctions may have impacted the company's ranking, but Iran's state-owned oil and gas company remains one of the top producers and suppliers of greenhouse gases.

Coal India: India's state-owned coal company is the world's single largest generator of coal-based pollution, slightly ahead of China's Shenhua Group, based on 2015 data. It is also the fourth largest producer of greenhouse gases among fossil fuel companies. As the dirtiest source of energy, coal burning contributes heavily to ocean acidification.

Shenhua Group: China's state-owned largest coal company was slightly behind its Indian counterpart, Coal India, in total greenhouse gas emissions. The company ranks fourth in greenhouse gas production among the world's top fossil fuel companies, and overall, China's coal industry is by far the world's biggest source of greenhouse gas production and supply—responsible for more than 14% of human-sourced greenhouse gas emissions between 1988 and 2015, according to CDP.

Coca-Cola Company: The Atlanta-based beverage behemoth is the top producer of plastic rubbish making its way into the world oceans, wreaking havoc on wildlife and sending plastic particles into the food chain, according to a global sampling last year of trash (Greenpeace and Break Free From Plastic) collected in 42 countries around the world. The world's leading beverage bottler, which markets hundreds of brands of carbonated beverages, water, and juices, said last year it uses three million tons of plastic packaging annually.

PepsiCo: A global sample of garbage collected on ocean shores and waters last year by Greenpeace and Break Free from Plastic organizations found that the North Carolina-based snack and beverage giant was second to Coca-Cola as the leading source of plastic garbage in the world's oceans. Plastic packaging can take hundreds of years to decompose, and in the oceans, it breaks down into tiny pieces that enter the food chain and kill wildlife that consumes it.

Nestlé: The Swiss multinational food and beverage company, whose brands include Gerber baby food, Perrier water, and KitKat chocolate among others, is the third largest source of plastic trash found in the oceans, according to a global sample of garbage collected on ocean shores last year by Greenpeace and Break Free From Plastic organizations. The company disclosed this year that it uses 1.7 million tons of plastic annually to produce its plastic packaging.

Danone: The French multinational food and beverage company, whose brands include Activia and Evian, is the fourth largest source of plastic garbage in the world's oceans. Danone disclosed last year that it uses about 750,000 tons of plastic annually to package its products. Plastic is the most common trash found in the oceans and it wreaks havoc on marine ecosystems.

Mondelez International: A global sample of ocean garbage collected last year found that the Illinois-based confectionary, food, and beverage company is the fifth most common source of plastic marine trash. Unlike other major plastic garbage producers on this list, the maker of Oreo cookies and Cadbury chocolate did not participate in recent U.N. program urging packaging producers to disclose the amount of plastic they use and to commit to more environmentally friendly practices in the future.

Agrochemical Companies: Bayer, following its 2018 acquisition of Monsanto (a genetically modified seed maker, and the world's largest genetic pesticide and herbicide (Roundup® etc. sales force) a German multinational pharmaceutical and chemical company became the world's largest supplier of seeds and agricultural chemicals. These chemicals are a key source of non-point water pollution, tragically with their glyphosate and other pollutants now found in all fresh and ocean waters globally. Like the oil that leaks from vehicles, pesticides like the ones produced by Bayer wind up in rivers that carry agricultural runoff into the oceans.

Syngenta: The Switzerland-based, Chinese-owned agrochemical and seeds producer was the world's largest by sales in 2017 before Bayer took first position following its acquisition of Monsanto in 2018. As the second largest agrochemical and seeds producer, Syngenta is one of the largest suppliers of herbicides and pesticides used in agricultural activity. These chemicals wash into rivers that carry them into coastal ecosystems.

BASF: The largest chemical company in the world by revenue is also a leading producer of agricultural pesticides, herbicides, and fertilizers. The German multinational company ranked third in agrochemical sales in 2017. Many of these chemicals wind up in the ocean through non-point water pollution, meaning there is no single specific source of contamination.

Corteva: DowDuPont finalized on June 3 the separation of its agricultural business unit into a standalone company, Corteva Agri sciences. The move comes after the completion of a merger between Michigan-based The Dow Chemical Company and Delaware's E.I. du Pont de Nemours in 2017. The merger made DowDuPont the fourth largest seller of agrochemicals, the segment now under the Corteva name. Herbicides, pesticides, and fertilizers pollute ocean water from rainwater runoff that makes its way to rivers, tributaries, and coastal wetlands.

ADAMA: The Israel-based, Chinese-owned company is a leading producer of agricultural chemicals and the world's largest producer of generic pesticides. It was acquired by China National Chemical Group in 2011. ChemChina later bought Syngenta, putting two major agrochemical producers under one parent company. Herbicides, pesticides, and fertilizers pollute oceans through rainwater runoff into rivers.

Meat and Dairy Corporations: JBS, the Brazilian multinational company is both the world's largest meat processor and the biggest single emitter of greenhouse gases in the global food industry, according to the Institute for Agriculture and Trade Policy and the international nonprofit GRAIN. The seller of meat, chicken, and pork products is responsible for more GHG emissions than the other top three leading meat and dairy companies combined. Carbon dioxide, the primary greenhouse gas, makes ocean water more acidic, which then causes great harm to marine ecosystems.

Tyson Foods: Arkansas-based Tyson is the second largest meat processor and the second biggest emitter of greenhouse gases in the global food industry, making it, according to the Institute for Agriculture and Trade Policy, one of the largest single sources of greenhouse gas emissions on the planet, when the energy used to produce animal feed and the agricultural chemicals used to grow it are considered. The primary greenhouse gas, carbon dioxide, makes seawater more acidic and harmful to ecosystems.

Cargill: The Minnesota-based, privately held conglomerate is heavily involved in the production of livestock and grains for livestock nutrition. Cargill's operations are not limited to meat and poultry as the company's diverse businesses include agriculture products, food and beverages, beauty products and more. The company ranks as the third biggest source of greenhouse gas emissions in the global livestock industry, according to a report last year by the Institute for Agriculture and Trade Policy and the international nonprofit GRAIN.

Dairy Farmers of America: The Kansas-based milk marketing cooperative is the fourth biggest source of greenhouse gas emissions in the meat and dairy industry if its livestock feed and agrochemicals operations are included. Much of the carbon dioxide from meat and dairy production winds up in the oceans, which makes them more acidic and harmful to sea life.

Fonterra: New Zealand commands an enormous share of the market in global whole milk powder, and much of that is under the control of Fonterra. The company is the fifth largest source of GHG emissions when the energy used to produce feedstock is included.

Our daily choices should include the best options such as sustainable small farms which work side by side with the nature and focus on local products and services.

## *Children Are the Future of the World. We Should Prepare Them to Choose Well*

How can we prepare our children for their future on a warming planet?

We must also consider their children, grandchildren and so on… This is looking far ahead, but that is what is needed now. Looking far ahead to the future will likely prevent ecologic collapse as well as acute and chronic economic, and even emotional loss.

We think it's better to enlighten children, students, and society, with the knowledge, perspicacious wisdom, ability, tools and skills needed now rather than to wait until humanity is forced to adapt, or runs out of time when it's too late. Apart from educating ourselves to consume less "stuff" in general, avoid meat and dairy, and to buy used or repurposed items, we know that learning to garden, self-reliance and self-sufficiency and so on is, in effect, "printing money". These assets will allow others to survive food shortages and social and economic collapse, and produce available resources to consume for themselves. Serious and widespread collapses are likely to occur during your lifetime, and your children's lifetimes.

Consequences of climate change include a die-off of most of the world's coral reefs, of biodiverse wet forests, risks to economic growth, and starvation-level food shortages, particularly in not yet developed countries. If a rapid reduction in heat-trapping gases does not occur soon enough, the world is likely to cross the 1.5 °C threshold roughly around 2040, when many of our youngest children will be in their 30s.

## *How to Prepare Children for a Warming and Acidic New World?*

Kids should be engaged in small conversations, which help them feel empowered rather than frightened, about the heating of our planet. As both a moral and a practical matter, they need to know that ocean warming and acidification is happening due to climate change, and that unfortunately it is likely to affect them and their children. To prepare children by talking with them, we suggest the need to start by studying available data, as well as oneself—self-reflection. For both children and adults, NASA's excellent climate website is very useful. Age-appropriate books and programs are important, and to hear about it from other children as well as adults. If we are able to educate our children about climate change, we will help them to make wise choices now and when they are adults. They will know, for one example, the risks of buying ocean-front property.

Please see Fig. 10.1 on some examples of houses constructed on top of dune vegetation. Sand dune vegetation protects land from ocean intrusion. If dune vegetation is removed, the roots that retain sediment responsible for the fixation and stability of the sand are also removed. If the original vegetation is replaced by concrete, the results are not pretty, and construction will likely collapse and fail.

Here we also show an example on Campeche beach, Florianópolis, Brazil, in 2020, of exactly what is not supposed to be happening at the beach. Bringing wood,

**Fig. 10.1** House constructed on top of dune vegetation. The dune vegetation is the protection of the land from ocean intrusion. Removal of dune vegetation, removes roots that retain sediment, responsible for the stability of the sand. If the original vegetation is replaced by concrete, the result is failure and pollution (and an ugly beachfront!)

steel and concrete, with plastic sandbags and not taking away the debris are not good solutions, because the ocean will likely to keep invading the now unstable sand-dune-foundation.

The illegal construction example pictured here, will surely end up in the ocean with toxic materials disrupting the entire ecological chain. Private owners want to stop the ocean and they had the approval of the local civil defense and the permit inspection agency!! The public power agency disapproved this construction, however with few personnel or necessary equipment, they do not come to inspect, stop or even to fine these infractions.

Our children need to have practical skills, which will be useful if disasters interrupt the normal flow of goods and services. Brief interruptions require knowing how to prepare for and survive a few days on your own, with flashlights, canned foods, and other emergency supplies. If longer interruptions occur, we all need to know how to garden, preserve food, safely gather and filter water, make repairs to a damaged home, sew clothing, provide medical care, install or fix solar arrays, use self-defense techniques, and so on. Children's interests should also guide the skills they learn. They can learn how to gather edibles, fish, and plant and preserve produce.

We should also prioritize our abilities to prepare for the projected climate impacts in each region; heat waves, poor air quality, sea level rise, and intense rainfall in the future, for example. We should learn and teach how to prevent heat stroke if one lives in a heat-wave area, including cities. It is important to explain how to stay safe in flash-flooding conditions, since intense precipitation is becoming more common. And unfortunately, because we all can expect worsening air quality, we should show them how to monitor outdoor air conditions.

Resilience in a climate crisis also emphasizes the importance of strengthening social networks within communities by encouraging them to create patterns of working together to overcome adversity, whether physical or psychological. We need to teach our children how to collaborate with and play well with others, while visualizing the best they can do for themselves and for others. This will help prepare them for difficult and harsh times, whether they arise from climate change or other natural or anthropic causes.

To engage in the citizenship of life, to the extent that we are able, to vote, write letters to 'the editor', contacting politicians or political leaders, and participate in community groups that are organizing others to do a project that will benefit the community. Our example becomes the model that our children need to build stronger communities. Also talking to them about how to work with others to overcome general or specific problems.

Policymakers at the U.S. Environmental Protection Agency (EPA) published a report in the Federal Register outlining how local communities should start planning for near-future catastrophes associated with climate change; and it is a must-read ifor yourself and for others to be prepared!

## *How to Plan for Natural Disaster Debris?*

We need to go beyond resilience to anticipate, plan, and prepare for impacts of climate change. In particular, it addresses 'how local communities can cope with debris and disaster following floods, hurricanes, or wildfires intensified by a changing climate, because it is expected to increase the frequency and intensity of some natural disasters', reads the report citing a 2014 National Climate Assessment. "The amount of debris generated by natural disasters, and the costs to manage it, will likely increase as a result", and "most threats from climate change are at 50 to 75 years out".

These threats represent an important change we have to address! Climate change is not some problem in the distant future. It is happening right now in every part of the world. When people say the increase size and scope of wildfires, hurricanes, and heat waves they are experiencing are unlike anything they have ever seen before, there is a reason for that. That reason is climate change. Also, the local pollution is another acute problem which decrease biodiversity as a whole (Fig. 10.2).

It is no secret that the world is seeing the consequences of climate change. Our shrinking glaciers, latitudinal shifting plant, and animal ranges are evidence that it is happening in real time, with past predictions now coming to fruition through loss of sea ice, intensified heatwaves, and sea level rise around the world.

Our EcoLogicProject (ELP) is confident that temperatures will continue rising for "decades to come", primarily due to human-caused greenhouse gas emissions. So, when it comes to planning, communities should assume "the worst-case scenario" as they adapt to the "debris-related impacts of climate change."

Across all nations, flooding is expected to intensify even in areas where total precipitation is projected to decline. Climate change is expected to increase the frequency and intensity of some natural disasters because we are going to a Pliocene like climate, the warmest periods of the earth.

Larger amounts of debris will affect wider areas, contributing to a greater risk of chemical and industrial release from facilities, and increased gas emissions from debris management activities, among other things.

**Fig. 10.2** Fauna inhabiting debris areas in the aquatic world

ELP has a mission to pinpoint the problems, include key stakeholders to identify potential debris streams due to sea level rise in coastal zones, evaluate potential sites of debris deposition and areas where they are exported in the marine world. Also to identify recycling programs to see if they can be scaled up during disasters, and address health and safety considerations specifically as it relates to future planning to mitigate and remediate the effects of climate change'.

So why use microfossils, and why study paleoclimate? What can past climate teach us about the future? We know by now that rising greenhouse gas concentrations are driving profound changes to the earth, including warming, sea level rise, increases in climate and weather extremes, ocean acidification, and ecological shifts.

Past climate variability, over the last centuries-to millennia, serves as the most relevant baseline against which to measure anthropogenic changes in climate. Farther back in time, larger reorganizations of the earth system provide critical information about the rates and magnitudes of physical, chemical, and ecological changes under a suite of different greenhouse gas concentrations.

The vast majority of instrumental observations of climate begin during the late twentieth century, during a period of rapid anthropogenic warming, and as such reflect the altered physics and chemistry of the anthropogenic era. The so-called paleoclimate records of past environmental changes capture a rich spectrum of Earth's history that climate scientists use to inform future climate changes in a variety of ways. Given that it will take many millennia for the earth system to come into equilibrium with present-day atmospheric greenhouse gas concentrations, past climate conditions help scientists understand the true sensitivity of the earth system to both small and large changes in climate conditions.

At their most basic level, records of past climate change serve as a critical backdrop for current anthropogenic climate trends, in many cases allowing for the separation of natural variability and greenhouse-induced trends in earth's climate. In recent millennia, atmospheric $CO_2$ concentrations were relatively stable, such that changes in solar irradiance and volcanic eruptions represented the primary drivers of global climate variability. During this time, global temperature varied by less than 0.5 °C and sea level varied by no more than 10 cm.

Exceptionally high-resolution records spanning the last several centuries allow for the quantification of past climate extremes such as severe drought, ENSO (el Nino southern Oscillation) events, wildfires, and tropical storms.

Indeed, the last millennia provides a wealth of data that climate scientists use to probe the relationship between the global climate state and the character of climate extremes. As such, it offers a rich testbed for climate models that are used to project twenty-first century climate changes, which must account for natural as well as solar, volcanic, and greenhouse-forced climate variability. Rising greenhouse gas concentrations reflect large-scale changes in the Earth's carbon cycle, such that, studies of past changes in carbon fluxes and associated climate changes are highly relevant to projections of future anthropogenic climate change.

Over the last million years, Earth has transitioned from glacial climate states characterized by markedly lower atmospheric $CO_2$ concentrations (200 parts per million) to interglacial climate states (280–300 parts per million) every

~100,000 years. Profound shifts in polar ice sheet mass, sea ice extent, sea level, ocean circulation, global temperature, precipitation patterns, and climate extremes accompanied these glacial-interglacial shifts.

Abundant evidence of abrupt climate change during the last glacial period illustrate the potential for rapid climate responses to much slower changes in climate forcing – a high-risk but highly uncertain scenario for twenty-first century climate changes.

Much further back in geologic time, deep-sea sediments record a climate state when changes in the Earth's carbon cycle caused atmospheric $CO_2$ concentrations to climb to ~1000 ppm—similar to levels expected in coming decades if emissions continue unabated. During this time, roughly 50 million years ago, global temperatures were as much as 8 °C warmer, sea level was more than 20 m higher, and ocean pH varied appreciably. While the rates of present-day atmospheric $CO_2$ change, temperature change, ocean pH change, and sea level rise are much higher than they were during past geologic intervals, these "hothouse" worlds hold key lessons for our climate future. This is especially true as models that were constructed and tested against instrumental climate data are charged with projecting climate changes that occur under vastly different conditions than today.

With our EcoLogicProject Climate Science Consultancy & Academic Advisory Service, we show ways to preserve and conserve the environment through global warming and climate change research, including effects of sea-level rise, and how to help communities to prepare now, for the near future.

ELP assesses and monitors the paleoenvironmental-change-record and its current-timeline implications. Collected data indicates carbon-overload, and acidification causing ocean warming, contributing to sea-level rise. This allows us to provide an accurate prediction of future impact effects.

Globally near-shore to deep ocean and in tropical to polar climates we research Forams microorganisms and their distribution in the sediment. Foraminifera are an accurate, inexpensive and easily handled proxy to identify at-risk environmental conditions, including areas experiencing hazardous pollution or contamination.

ELP offers unique access to a dedicated team of professionals who specialize in a variety of interrelated expertise. Our combined abilities solve multifaceted environmental challenges in an efficient, timely, and cost-effective manner.

ELP's cofounders comprise our core team. ELP's Science Division also collaborates with a diverse group of highly skilled environmental science colleagues and professionals with expertise in the the following fields: Marine biology, Marine geology & geophysics, Physical oceanography, Biological oceanography, Fish ecology, Environmental impacts, Endangered, exotic and invasive species, Marine mammals, birds, and turtles, Biostatistics, Marine habitat restoration, GIS, Coastal zone management, Underwater photography, Remote sensing data interpretation.

Our mission is committed to offering the best and most efficient oceanographic, coastal and environmental consulting services in the industry. Offering cost-effective solutions, we navigate among interrelated disciplines including physical oceanography, hydrology, marine biology, coastal & marine geology and climate science.

**Achieving accurate research results, we also manage, coordinate, and facilitate ocean research and education programs. ELP shares ocean exploration and observing, analysis, and prediction information through mentoring, advocacy, program management, and facilitation across a wide spectrum of communities. These include oceanographic and academia, formal and informal education, technological and industrial institutions, government entities and municipalities, and the public. Our logos represent what we fight for.**

# Concluding Remarks

## The Anthropocene Sixth Extinction: Prediction

First, we want to emphasize that the authors and the EcoLogicProject are in no way associated or affiliated with any political group or agenda, or social movement. Rather we are dedicated to the promotion of ecological sustainability, and to encouraging others to embrace an optimum climax biodiversity for ourselves, and all life on earth including future generations.

In these concluding remarks, we will repeat some topics, subjects and themes. This is to firmly emblazon the gravity of magnitude regarding these topics, and firmly imprint upon the psyche of the reader the responsibility of ecological stewardship that inherently we all hold, individually and collectively.

Our most basic needs; clean water, air and soil are the sustenance of our lives. These are the only things that give us true stability and everything we need to survive and flourish.

Here again we wish to discuss some history, segueing into our current human environmental impact effects.

"Don't let a good crisis go to waste". This 'crisis platform' appears to be the unofficial anthem of 'special interest' for-profit—politicians-for-hire. These clever con-artists promise greater "security" from whatever threat—real or imagined. Although when carefully scrutinized these 'promises' are all too often simply man-made laws legislated to benefit—the-few; while—the-many (who may feel safer), actually experience a greater cost—loss of freedoms. All the while—the-few, grossly ignore nature's greater laws of sustainability, while greedily feeding at the trough of power, control and profit-at -any expense.

We observe this "crisis" platform taking advantage of virus, plagues and pandemics, which historically have coincided with global climate changes, and not surprisingly have caused governance power vacuums where stronger forces take advantage of weakened social and economic conditions.



P. P. Beck Eichler, C. P. Barker, *Benthic Foraminiferal Ecology*,
https://doi.org/10.1007/978-3-030-61463-8

Despair, paranoia and hysteria promulgated by political agendas and media influences only create division, and as we may well remember, 'Divide and Conquer' is as old as politics and war.

There are two ways to lead people. One way is to divide them by creating a mutual enemy (virus, social threats, etc.), and lead a crusade to conquer the threat. A life or death urgency unleashes unconditional support with a 'winning at-any-cost' mentality. Another way to lead is by uniting people with a shared ambition that inspires people to build together. Creating a life quest to build something bigger unites people.

In June of 2020, amidst Black Lives Matter protest riots raging across the United States, during the Coronavirus (COVID-19) pandemic/diversion, the political agenda continued to dominate as US President Donald Trump signed an executive order to remove environmental checks, reducing environmental protections in order to 'fast-track the nation's economic recovery'. This blindingly selfish glut-greed mentality prioritizes profit over reason and accountability, even as irrefutably empirically accurate scientific data proves that globally our human activity is rapidly affecting a potential sixth mass extinction.

Looking back, we see that in 541 AD the "Justinian Plague" followed a mini ice age after volcanic eruptions a decade earlier. In 1347 AD, a couple of decades after another global climate event, we had a Bubonic plague called "The Black Death" which weakened Europe creating power vacuums in which, for example, the "Arab Conquest" Islamic Caliphate "land grab" was experienced throughout Iberia, Spain and Portugal, and then moving through France and Eastward.

The resulting plague induced huge population collapses gave rise to the Turks Ottoman Empire where for four centuries two million slaves were taken from Russia, Belarus, Ukraine, and Poland, for a period of about two centuries. These slaves were taken and sold to the Ottomans. Let us keep in mind, for perspective, that at around the year 1600 AD., there were only about 1.6 million people in Russia.

If history teaches us anything, it is how to react to and prepare for crisis rather than being doomed to repeating mistakes. In terms of preparing ourselves, what should we do with what we have learned? Do we want to be polarized in times of crisis or do we want to have unity? Do we want cultural changes imposed upon us, or do we want to make personal and social changes of our own accord?

"It is the mark of an educated mind to be able to entertain a thought without accepting it."—Aristotle.

Divisions are an illusion for both the leader and their followers; they create a fictional reality. Creating a common enemy is the most usual way to divide people—leaders force us to choose between being in favor of or against their crusade.

The Unite & Build Approach is also as old as humanity. "Join me and follow the mission, not me" is an ancient Taoist wisdom resulting in making the world a better place. The ultimate objective of 'unite and build' leadership is not gaining power, but empowering people to achieve something bigger through collaboration, not competition.

Promoting unity does not mean brainwashing people or embracing conformity. Groupthink is a dangerous behavior that eliminates conflict to reach consensus—it lowers the ability to critically evaluate alternatives. Things are changing now in terms of what we want personally, and what kind of society and natural world we wish to live in.

Our oceans, lakes and rivers are increasingly empty of life as a result of humanities greedy little-mindedness. Research suggests we are fast approaching disastrous effects of this sixth Anthropocene extinction. However, we can successfully surmount the challenges of biodiversity loss and climate change, and dramatically alter the trajectory for what R. Buckminster Fuller called "Spaceship Earth". His concept is for a comprehensive planetary planning that resulted in new strategies intended to enable all of humanity to live with freedom, comfort and dignity, without negatively affecting the earth's ecosystems or regenerative ability.

## A Planet Currently "In Crisis"

Every day new studies show a growing body of evidence demonstrating widespread ecological collapse and biodiversity loss. We know that as average temperatures rise and the frequency of extremely warm years increases, the impacts of habitat loss and fragmentation become increasingly apparent.

Biodiversity is the global food system, yet agricultural and industrial expansion has led to the loss of over 85% of wetlands, altered 75% of land surface, and affected 66% of ocean area. In the United States, agricultural land makes up nearly 45% of the land area, illustrating the important role working landscapes could play in mitigating the worst effects of biodiversity loss, while adapting to the extreme weather patterns and climate trends we presently face. Yet in 2018, a report from the Institute for Agriculture and Trade Policy found that less than half of states in the U.S. have climate adaptation plans that include agriculture; even fewer give attention to how farms and fields can be managed as part of the larger landscape.

Where government climate adaptation plans fail to account for agriculture, private sector actors are blazing new trails. At the Rodale Institute, for example, Dr. Bronner's and Patagonia launched the Regenerative Organic Certification to provide farmers with a framework for increasing soil health and promoting biodiversity while improving animal welfare and ensuring fair labor conditions.

Leading-edge integrated multi-trophic aquaculture (IMTA) advanced by AquaVitae, a trans-Atlantic aquaculture working group, is experimenting with co-cultured species throughout the water column to mimic the ecology of natural systems. As of 2018, over 550 active payments for ecosystem service (PES) programs around the world pay farmers, suppliers and land managers an estimated $42 billion per year to ensure ecosystem services, such as pollination or water filtration. This is just the beginning of a global effort, to accurately account for undervalued biological functions of agricultural production systems, to close loops, and to leverage these services to regenerate ecosystems. The aforementioned efforts, while noble, are not enough. Studies find that a sixth mass extinction event is accelerating at an unprecedented rate.

A 2015 Stanford University team scientists led by biologist Paul Ehrlich published a study claiming we were entering a sixth mass extinction event triggered by the rapid decline in biodiversity. An update has found the rate of extinctions is increasing at an unprecedented rate, with scientists fearful that it could kick off a domino effect posing a real threat to human existence.

Humankind relies heavily on biodiversity and healthy ecosystems to survive and flourish, whether that's through the bees that pollinate our crops, marine species that keep the oceans healthy, dependable sources of food and water or the ever-expanding array of nature-sourced lifesaving medicines. As species disappear due to habitat destruction, wildlife trade, pollution and climate change, so do the essential parts of a delicate-interdependent system we all depend on, consciously or not, for our very life and well-being.

"When humanity exterminates populations and species of other creatures, it is sawing off the limb on which it is sitting, destroying working parts of our own life-support system," says Ehrlich. "The conservation of endangered species should be elevated to a national and global emergency for governments and institutions, equal to climate disruption to which it is linked."

Ehrlich's international team of researchers examined the distribution of critically endangered species around the globe, finding that 515 of them, 1.7% of the total analyzed, are on the brink of extinction and are likely to disappear in the next two decades. They estimate that around the same number went extinct across the entire twentieth century. The 515 species considered to be on the brink of extinction had fewer than 1000 individuals left, with around half of those having less than 250. The majority of these species inhabit environmentally delicate tropical and subtropical areas susceptible to human impacts. This refers to a situation where the extinction of one species influences other species that rely on it for survival, thereby also placing them at the already mentioned 'domino effect' risk of extinction as part of a destructive chain reaction.

Researchers propose a number of ways in which we can immediately address the problem, including but definitely not limited to a global ban on the trade of wild species. Also with the cutting and burning of vast wild-lands, such as the Amazon and Atlantic forests, the study highlights species and regions where conservation resources can be best directed, which in turn can highlight which factors are most impactful in driving these accelerating extinction rates.

"The links between human health and wellbeing, and the health of our planet are well known," says Dr. Rohan Clarke, a lecturer in the School of Biological Sciences at Australia's Monash University. "This research highlights the fragility of the Earth's support systems and the urgent need to act". The call for the conservation of endangered species has to be elevated to a national and global emergency is both warranted and urgent!"

Entire ecosystem changes can take the whole community to extinction. If history teaches us anything, it is how to react to and prepare for crisis rather than repeat mistakes. Research suggests we are fast approaching disastrous effects of this sixth Anthropocene extinction. However, we can successfully surmount the challenges of biodiversity loss and climate change and dramatically alter the trajectory if we can pinpoint and remediate problems within a near future. With our planet "in crisis", evidence demonstrates widespread ecological collapse and biodiversity loss. We know that as average temperatures rise and the frequency of extremely warm years increases, the impacts of habitat loss and fragmentation become even more increasingly apparent.

We are with without-a-doubt entering a sixth mass extinction event because of the extremely rapid decline in biodiversity. The majority of these species inhabit environmentally delicate tropical and subtropical areas and are highly susceptible to human impacts. To be very clear; this refers to the extinction of one species impacting other species that rely on it for survival, placing them at a 'domino effect' risk of extinction as part of a destructive chain reaction.

In 2020, the Intergovernmental Science-Policy Platform on Biodiversity and Ecosystem Services reported that the current rate of extinction is "tens to hundreds of times higher than average over the past 10 million years — and it's accelerating." This year, biodiversity loss topped the World Economic Forum's annual Global Risk Report for "Likely" and "Impactful" long-term risks.

With food security, supply chain stability, and our EcoLogicProject vision for sustainable development at risk, it is integral that we restore and reconnect our landscapes and watersheds, reduce-to-elimination of emissions, sequester carbon, and steward biodiversity. Stop cutting and burning forests, stop global trade of wild species, study and protect, preserve and conserve our planet's biodiversity.

EcoLogicProject.com shares ocean exploration and observing, analysis, and prediction information through mentoring, advocacy, program management, and facilitation across a wide spectrum of communities. These include oceanographic and academia, formal and informal education, technological and industrial institutions, government entities and municipalities, and the public.

# References


Aiello IW, Bova SC, Holbourn AE, Kulhanek DK, Ravelo AC Rosenthal Y (2019) Climate, sea level and tectonic controls on sediment discharge from the Sepik River, Papua New Guinea during the Mid-to Late Pleistocene. Marine Geology 415:105954

Altenbach AV, Lutze GF, Schiebel R, Schönfeld J (2003) Impact of interrelated and interdependent ecological controls on benthic foraminifera: an example from the Gulf of Guinea. Palaeogeogr Palaeoclimatol Palaeoecol 197(3–4):213–238

Alve E (1991) Benthic foraminifera in sediment cores reflecting heavy metal pollution in Sorfjord, Western Norway. J Foraminifer Res 21(1):1–19

Alve E (1995) Benthic foraminiferal responses to estuarine pollution: a review. J Foraminifer Res 25(3):190–203

Alley RB (2000) Ice-core evidence of abrupt climate changes. Proceedings of the National Academy of Sciences 97(4):1331–1334

Andersson C (1997) Transfer function vs. modern analog technique for estimating Pliocene Sea-surface temperatures based on planktic foraminiferal data, western equatorial Pacific Ocean. J Foraminifer Res 27(2):123–132

Araújo P, Amaral R (2016) Mapping of coral reefs in the continental shelf of Brazilian northeast through remote sensing. Rev Gestão Cost Integr 16:5

Araújo HAB, De Jesus Machado A (2008) Benthic foraminifera associated with the South Bahia coral reefs, Brazil. J Foraminifer Res 38(1):23–38

AWH B, Tolderlund DS (1971) Distribution and ecology of living planktonic foraminifera in surface waters of the Atlantic and Indian Oceans. Micropaleontology of the Oceans, pp. 105–150

Azevedo CAA, Carneiro MAA, Oliveira SR, Marinho-Soriano E (2011) Macrolgae as an indicator of the environmental health of the Pirangi reefs, Rio Grande do Norte, Brazil. Rev Bras Farm 21(2):323–328

Baker RD, Hallock P, Moses EF, Williams DE, Ramirez A (2009) Larger foraminifers of the Florida reef tract, USA: distribution patterns on reef-rubble habitats. J Foraminifer Res 39:267–277

Baptista-Neto JB, Smith BJ, McAllister JJ (2000) Heavy metal concentrations in surface sediments in a nearshore environment, Jurujuba sound, Southeast Brazil. Environ Pollut 109(1):1–9

Barbosa CF, Oliveira-Silva P, Seoane JCS, Cordeiro RC, Ferreira BP, Soares-Gomes A, de Almeida CM, Queiroz GP, Lamana T, Lisboa S, Duran D (2006) Foraminifera on coral reefs of Brazil: the FOCO project, vol 29. Anuário do Instituto de Geociências, Rio de Janeiro, pp 526–527

Barbosa CF, Prazeres MF, Ferreira BP, Seoane JCS (2009) Foraminiferal assemblage and reef check census in coral reef health monitoring of east Brazilian margin. Mar Micropaleontol 73:62–69




P. P. Beck Eichler, C. P. Barker, *Benthic Foraminiferal Ecology*,
https://doi.org/10.1007/978-3-030-61463-8

Barbosa CF, Ferreira BP, Seoane JCS, Oliveira-Silva P, Gaspar ALB, Cordeiro RC, Soares-Gomes A (2012) Foraminifer-based coral reef health assessment for southwestern Atlantic offshore archipelagos, Brazil. J Foraminifer Res 42:169–183

Batsch AJGK (1791) Tabula affinitatum regni vegetabilis: quam delineavit, et nunc ulterius adumbratam tradit. in bibliopolio, vulgo Landes-Industrie Comptoir

Berger WH, Wefer G (1991) Productivity of the glacial ocean: discussion of the iron hypothesis. Limnol Oceanogr 36(8):1899–1918

Bicchi E, Debenay JP, Pagès J (2002) Relationship between benthic foraminiferal assemblages and environmental factors in atoll lagoons of the central Tuamotu archipelago (French Polynesia). Coral Reefs 21:275–290

Bijma J, Erez J, Hemleben C (1990a) Lunar and semi-lunar reproductive cycles in some spinose planktonic foraminifers. J Foraminifer Res 20(2):117–127

Bijma J, Faber WW, Hemleben C (1990b) Temperature and salinity limits for growth and survival of some planktonic foraminifers in laboratory cultures. J Foraminifer Res 20(2):95–116

Billups K, Eichler PPB, Vital H (2020) Sensitivity of benthic foraminifera to carbon flux in the western tropical Pacific Ocean. J Foraminifer Res 50(2):235–247. https://doi.org/10.2113/gsjfr.50.2.235

Boltovskoy E (1973) Daily vertical migration and absolute abundance of living planktonic foraminifera. J Foraminifer Res 3(2):89–94

Boltovskoy E, Giussani G, Watanabe S, Wright R (1980) Atlas of benthic shelf foraminifera of southwest atlantic. Dr. W. Junk, The Hague, 174 p

Boltovskoy E, Scott DB, Medioli FS (1991) Morphological variations of benthic foraminiferal tests in response to changes in ecological parameters: a review. J Paleo 65:175–185

Boudreau C, Heath E, Seuntjens J, Ballivy O, Parker W (2005) IMRT head and neck treatment planning with a commercially available Monte Carlo based planning system. Phys Med Biol 50(5):879

Bradshaw JS (1959) Ecology of living planktonic foraminifera in the north and equatorial Pacific Ocean. Contrib Cushman Found Foram Res 10:25–64

Brady HB (1881) Notes on some of the reticularian Rhizopoda of the" Challenger" Expedition. Part III. 1. Classification. 2. Further notes on new species. 3. Note on Biloculina mud. Quarterly Journal of Microscopical Science, new series, 21:31–71

Brady HB (1884) Report on foraminifera, vol 9. HM Stationery Office, London

Brönnimann P (1979a) Recent benthonic foraminifera from Brazil-Morphology and Ecology: Part 4. Trochaminids from the shelf with description of Paratrochammine n. gen. Paleontol Zh 63:5–25

Brönnimann P (1979b) Recent benthonic foraminifera from Brasil. morphology and ecology. Part IV: Trochaminids from the Campos Shelf with description of Paratrochammina n gen. Paleontol Zh 53(1/2):5–25

Buhl-Mortensen L, Mortensen PB (2004) Symbiosis in deepwater corals. Symbiosis 37:33–61

Burone L, Pires-Vanin AMS (2006) Foraminiferal assemblages in the Ubatuba Bay, south-eastern Brazilian Coast. Scientia Marina 70(2):203–217

Buttermore RE, Turner E, Merrill MG (1994) The introduced northern Pacific seastar Asterias amurensisin Tasmania. Mem Qld Museum 36:1–25

Campos EJD, Ikeda Y, Castro BM, Gaeta SA, Lorenzzetti JA, Stevenson MR (1996) Experiment studies circulation in the Western South Atlantic. EOS Trans Am Geophys Union 77(27):253–259

Cardoso BPK (2000) Caracterização dos foraminíferos no Channel de São Sebastião (SP) e, sua utilização como indicadores de massa d'água e de algumas variáveis ambientais. Dissertação de Mestrado. Instituto Oceanográfico da Universidade de São Paulo

Carilli J, Walsh S (2012) Benthic foraminiferal assemblages from Kiritimati (Christmas) island indicate increased nitrification has occurred on a decadal scale. Mar Ecol Prog Ser 456:87–99

Carlton JT (1985) Transoceanic and interoceanic dispersal ofcoastal marine organisms: the biology of ballast water. Oceanogr Mar Biol Annu Rev 23:313–371

Carlton JT, Geller JB (1993) Ecological roulette: the global transport of nonindigenous marine organisms. Science 261:78–82

Carlton JT, Reid DM, Van Leeuwen H (1995) Shipping study. The role of shipping in the introduction of non-indigenous aquatic organisms to the coastal waters of the United States (other than the Great Lakes) and analysis of control options. The National Sea Grant College Program/ Connecticut Sea Grant ProjectR/ES-6. Department of Transportation, United States Coast Guard. Washington, DC and Groton, Connecticut. Report CG-D-11-95.Government Accession Number AD-A294809 (213 p) and Appendices A-I (122 p)

Carnahan EA, Hoare MM, Hallock P, Lidz BH, Reich CD (2009) Foraminiferal assemblages in Biscayne Bay, Florida, USA: responses to urban and agricultural influence in a subtropical estuary. Mar Pollut Bull 59:221–233

Castro BD, Miranda LD (1998) Physical oceanography of the western Atlantic continental shelf located between 4 N and 34 S. The sea 11(1):209–251

Castro CB, Pires DO (2001) Brazilian coral reefs: what we already know and what is still missing. Bull Mar Sci 69:357–371

Center For The Inland Bays (1995) Online. https://www.inlandbays.org/cib-pm/messages.php

Chapman F (1915) Report on the foraminifera and Ostracoda obtained by the FIS "Endeavour" from the east coast of Tasmania, and off cape wiles, South Australia

Charman D (2002) Peatlands and environmental change. John Wiley & Sons Ltd

Charman DJ, Hendon D (2000) Long-Term Changes in Soil Water Tables over thePast 4500 Years: Relationships with Climate and NorthAtlantic Atmospheric Circulation and Sea Surface Temperature. Climatic Change 47(1–2):45–59

Chu KH, Tam PF, Fung CH, Chen QC (1997) A biological survey of ballast water in container ships entering Hong Kong. Hydrobiologia 352:201–206

Citizen's, P.W.S.R., ADVISORY COUNCIL (2004) Coping with Technological Disasters: A User Friendly Guidebook, Anchorage

Clarke KR (1993) Non-parametric multivariate analysis of changes in community structure. Aust J Ecol 18:117–143

Clarke O (2020). https://www.osborneclarke.com/insights/uk-coronavirus-act-overview/

Clarke KR, Ainsworth M (1993) A method of linking multivariate community structure to environmental variables. Mar Ecol Prog Ser 92:205–219

Clarke KR, Warwick RM (1994a) Changes in marine communities: an approach to statistical analyses and interpretation. Natural Environment Research Council, Plymouth

Clarke KR, Warwick RM (1994b) Change in marine communities: an approach to statistical analysis and interpretation. Plymouth Marine Laboratory, Plymouth, p 144

Cockey EM, Hallock PM, Lidz BH (1996) Decadalscale changes in benthic foraminiferal assemblages off key largo, Florida. Coral Reefs 15:237–248

Cohen AN, Weinstein A, Emmett MA, Lau W, Carlton JT (2001) Investigations into the introduction ofnon-indigenous marine organisms via the cross-continental trade inmarine baitworms. A report for the U.S. Fish and Wildlife Service, San Francisco Bay Program, Sacramento CA. San Francisco Estuary Institute, Richmond, CA, 28 p

Collins ES, Scott DB, Gayes PT, Medioli FS (1995) Foraminifera in Winyah Bay and north inlet marshes, South Carolina; relationship to local pollution sources. J Foraminifer Res 25(3):212–223

Conan SH, Ivanova EM, Brummer GJ (2002) Quantifying carbonate dissolution and calibration of foraminiferal dissolution indices in the Somali Basin. Mar Geol 182(3–4):325–349

Cushman JA (1922) Shallow-water foraminifera of the Tortugas region (no. 311). Carnegie Institute, Pittsburgh, PA

D'Orbigny AD (1826) Tableau méthodique de la classe des Céphalopodes. Ann Scis Nat 1:96–314

Debenay JP, Eichler BB, Guillou J-J, Eichler-Coelho P, Coelho C, Porto-Filho E (1997) Comportement des peuplements de foraminifères etcomparaison avec l'avifaune dans une lagune strongly stratifiée:la Lagoa da Conceição (S.C., Brésil). Rev Paléobiol Genève 16(1):55–75

Defrance JLM (1822) Larvaire, vol 25. Dictionnaire des Sciences Naturelles, Paris, pp 287–288

Diaz RJ (2001) Overview of hypoxia around the world. J Environ Qual 30(2):275–281

Dobbs FC, Diallo AA, Doblin MA, Drake LA, Graczyk T, Jiang X, Johengen TH, Kampschmidt L, Rublee P, Reid DF, Ruiz G (2003) Pathogens in ships' ballast water and sediment residuals. In: Third International Conference on Marine Bio invasions, San Diego, CA. thewebat: http://www.sgmeet.com/mb/viewabstract2.asp?AbstractID=77andSessionID=20

Ehrenberg CG (1839) Uber die Bildung der Kreidefelsen und des Kreidemergels durch unsichtbare Organismen. Abhandlungen der Koniglichen Akademie der Wissenschaften zu Berlin 1838:59–147

Ehrlich P (2015). https://news.stanford.edu/2020/06/01/loss-land-based-vertebrates-accelerating/

Eichler BB, Debenay JP, Bonetti C, Duleba W (1995) Répartition des Foraminifères benthiques dans la zone Sud-Ouest du Système laguno-estuarien d'Iguape-Cananéia (Brésil). Bol Inst Oceanogr 43(1):1–17

Eichler BB, Eichler PB, Miranda LB, Bergamo AL, Bernardes MEC, Pereira ERM, Kfouri PBP, Pimenta FM (2001) Use of foraminifera as bioindicators of marine influence in Guanabara Bay (RJ, Brazil). Res J Geosci 28(2):251–262

Eichler PP, Eichler BB, de Miranda LB, Pereira EDR, Kfouri PB, Pimenta FM, Bérgamo AL, Vilela CG (2003) Benthic foraminiferal response to variations in temperature, salinity, dissolved oxygen and organic carbon, in the Guanabara Bay, Rio de Janeiro, Brazil. Anu Inst Geocienc 26:36–51

Eichler PPB, Castelão GP, Pimenta FM, Eichler BB, De Miranda LB, Rodrigues AR, Pereira ER (2004) Foraminifera and thecamoebians as indicator of hydrodynamic process in a choked coastal lagoon, Laguna estuarine system, SC, Brazil. J Coast Res 39:1144–1148

Eichler PPB, Eichler BB, de Miranda LB, Rodrigues AR (2007) Modern foraminiferal facies in a subtropical estuarine channel, Bertioga, São Paulo, Brazil. J Foraminifer Res 37(3):234–247

Eichler PPB, Gupta BS, Eichler BB, Braga ES, Campos EJ (2008) Benthic foraminiferal assemblages of the South Brazil: relationship to water masses and nutrient distributions. Cont Shelf Res 28(13):1674–1686

Eichler PP, Billups K, Cardona CCV (2010) Investigating faunal and geochemical methods for tracing salinity in an Atlantic coastal lagoon, Delaware, USA. J Foraminifer Res 40(1):16–35

Eichler PP, Eichler BB, Gupta BS, Rodrigues AR (2012a) Foraminifera as indicators of marine pollutant contamination on the inner continental shelf of southern Brazil. Mar Pollut Bull 64(1):22–30

Eichler PPB, Rodrigues AR, Eichler BB, Braga ES, Campos EJD (2012b) Tracing latitudinal gradient, river discharge and water masses along the subtropical south American coast using benthic foraminifera assemblages. Braz J Biol 72:723–759

Eichler PP, Pimenta FM, Eichler BB, Vital H (2016) Living benthic foraminiferal species as indicators of cold-warm water masses interaction and upwelling areas. Cont Shelf Res 116:116–121

Eichler PPB, de Farias CL, Amorin A, De Moura DS, Andrade APU, Corrêa AA, Martins JF, Vital H, Gomes MP (2019a) Symbiont-bearing Foraminifera from Reefal areas: a case study from Rio Grande Do Norte (RN, Brazil). J Foraminifer Res 49(2):131–140. https://doi.org/10.2113/gsjfr.49.2.131

Eichler PP, Rodrigues AR, Gomes MP, Vital H (2019b) Benthic foraminifera as indicators of river discharge in the Western South Atlantic continental shelf. Mar Geol 415:105973

Eichler-Coelho PB (1996) Estudo das associações de foraminíferos e tecamebas da região estuarino-lagunar de Cananéia-Iguape (SP), e sua aplicação na determinação do impacto ecológico do Valo Grande. Dissertação de Mestrado. Instituto Oceanográfico, Universidade de São Paulo, 106 p

Embley RW, Chadwick WW, Baker ET, Butterfield DA, Resing JA, De Ronde CE, Tunnicliffe V, Lupton JE, Juniper SK, Rubin KH Stern RJ (2006) Long-term eruptive activity at a submarine arc volcano. Nature 441(7092):494–497

Emiliani C (1954a) Depth habitats of some species of pelagic foraminifera as indicated by oxygen isotope ratios. Am J Sci 252(3):149–158

Emiliani C (1954b) Temperatures of Pacific bottom waters and polar superficial waters during the tertiary. Science 119(3103):853–855

Emiliani C (1956) Oligocene and Miocene temperatures of the equatorial and subtropical Atlantic Ocean. J Geol 64(3):281–288
Emiliani C (1966) Paleotemperature analysis of Caribbean cores P6304-8 and P6304-9 and a generalized temperature curve for the past 425,000 years. The Journal of Geology 74(2):109–124
Ericson DB (1959) Coiling direction of Globigerina pachyderma as a climatic index. Science 130(3369):219–220
Ewing, Maurice, D. B. Ericson, BC Heezen, JL Worzel, Goesta W (1954) Exploration of the deep-sea floor 145–168
Fedorov AV, Brierley CM, Emanuel K (2010) Tropical cyclones and permanent El Niño in the early Pliocene epoch. Nature 463(7284):1066–1070
Ferreira BP, Maida M (2006) Monitoramento dos recifes de coral do Brasil. MMA, Secretaria de Biodiversidade e Florestas, Brasília, DF
Fichtel LV, Moll JV (1798) Testacea microscopica aliaque minuta ex generibus Argonauta et Nautilus ad naturam delineata et descripta, vol 12. Anton Pichler, Vienna, pp 1–123
Flood PG, Orme GR (1988) Chapter 6 mixed siliciclastic/carbonate sediments of the northern great barrier reef province, Australia. Dev Sedimentol 42:175–205
Fujita K (2004) A field colonization experiment on small-scale distributions of algal symbiont-bearing larger foraminifera on reef rubble. J Foraminifer Res 34:169–179
Furtado VV (1995) Sedimentação quaternária no Canal de São Sebastião. Publção Esp Inst Oceanogr. São Paulo 11:27–35
Gehrels WR (1994) Determining relative sea-level change from salt-marsh foraminifera and plant zones on the coast of Maine, U.S.A. J Coast Res 10(4):990–1009. www.jstor.org/stable/4298291
Giraudeau J (1993) Planktonic foraminiferal assemblages in surface sediments from the southwest African continental margin. Marine Geology 110(1–2):47–62
Gollasch S, Lenz J, Dammer M, Andres HG (2000) Survival of tropical ballast water organisms during a cruise from the Indian Ocean to the North Sea. Journal of Plankton Research 22(5):923–937
Gollasch S, Macdonald E, Belson S, Botnen H, Christensen JT, Hamer JP, Houvenaghel G, Jelmert A, Lucas I, Masson D, McCollin T, Olenin S, Persson A, Wallentinus I, Wetsteyn LPMJ, Wittling T (2002) Life in ballast tanks. In: Leppakoski E, Gollash S, Olenin S (eds) Invasive species of Europe-distribution, impact and management. Kluwer, Dordrecht, pp 217–231
Gomes MP, Vital H, Stattegger K, Schwarzer K (2016) Bedrock control on the Assu incised valley morphology and sedimentation in the Brazilian equatorial shelf. Int J Sedim Res 31:181–193
Gupta AK, Anderson DM, Overpeck JT (2003) Abrupt changes in the Asian southwest monsoon during the Holocene and their links to the North Atlantic Ocean. Nature 421(6921):354–357
Hallock P (2012) July The FoRAM Index revisited: uses, challenges, and limitations. In Proceedings of the 12th International Coral Reef Symposium, Cairns, Australia (pp. 9–13)
Hallock P (2000) Larger foraminifera as indicators of coral-reef vitality. In: Martin RE (ed) Environmental micropaleontology. Kluwer Academic/Plenum, New York, pp 121–150
Hallock P (2012) The FORAM index revisited. Uses, challenges, and limitations. In: Proceedings of the 12th international coral reef symposium
Hallock P, Forward LB, Hansen HJ (1986) Influence of environment on the test shape of Amphistegina. J Foraminifer Res 16:224–231
Hallock P, Lidz BH, Cockey-Burkhard EM, Donnelly KB (2003) Foraminifera as bioindicators in coral reef assessment and monitoring: the FORAM index. Environ Monit Assess 81 (1–3):221–238
Harari J, Camargo RD (1998) Numerical modeling of the coast region of Santos (SP): tidal circulation. Rev Bras Oceanogr 46(2):135–156
Haymon RM, Fornari DJ, Von Damm KL, Lilley MD, Perfit MR, Edmond JM, Shanks III WC, Lutz RA, Grebmeier JM, Carbotte S, Wright D (1993) Volcanic eruption of the mid-ocean ridge along the East Pacific Rise crest at 9 45–52′ N: Direct submersible observations of seafloor phenomena associated with an eruption event in April, 1991. Earth and Planetary Science Letters 119(1–2):85–101

Hays JD, Imbrie J, Shackleton NJ (1976) Variations in the Earth's orbit: pacemaker of the ice ages. Science 194(4270):1121–1132

Hemleben C, Spindler M, Anderson OR (2012) Modern planktonic foraminifera. Springer Science and Business Media, Berlin

Hohenegger J (1994) Distribution of living larger foraminifera NW of Sesoko-Jima, Okinawa, Japan. Mar Ecol 15:291–334

Holbourn A, Kuhnt W, Schulz M, Erlenkeuser H (2005) Impacts of orbital forcing and atmospheric carbon dioxide on Miocene ice-sheet expansion. Nature 438(7067):483–487

Ikeya N (1977) Ecology of foraminifer in the Hamana Lake region on the Pacific coast of Japan. Rep Fac Sci Shizuoka Univ 11:131–159

Ingle JC (1977) Summary of late neogene planktic foraminiferal biofacies, biostratigraphy, and paleoceanography of the marginal north pacific ocean

Jones RW, Brady HB (1994) The challenger foraminifera. Oxford University Press, New York

Jorissen FJ, Fontanier C, Thomas E (2007) Chapter seven paleoceanographical proxies based on deep-sea benthic foraminiferal assemblage characteristics. Dev Mar Geol 1:263–325

Karpas RM (1978) The hydrography of Indian River and Rehoboth, Delaware's small bays. Unpublished Master Thesis, University of Delaware, 178 p

Katz ME, Katz DR, Wright JD, Miller KG, Pak DK, Shackleton NJ, Thomas E (2003) Early Cenozoic benthic foraminiferal isotopes: species reliability and interspecies correction factors. Paleoceanography 18(2)

Kawahata H, Nishimura A, Gagan MK (2002) Seasonal change in foraminiferal production in the western equatorial Pacific warm pool: evidence from sediment trap experiments. Deep-Sea Res II Top Stud Oceanogr 49(13–14):2783–2800

Kelmo F, Hallock P (2013) Responses of foraminiferal assemblages to ENSO climate patterns on bank reefs of northern Bahia, Brazil: A 17-year record. Ecological Indicators 30:148–157

Kjerfve B, Ribeiro CH, Dias GT, Filippo AM, Da Silva Quaresma V (1997) Oceanographic characteristics of an impacted coastal bay: Baía de Guanabara, Rio de Janeiro, Brazil. Cont Shelf Res 17(13):1609–1643

Kramer KJ, Botterweg J (1991) Aquatic biological early warning systems: an overview. In: Bioindicators and environmental management. Academic Press, London, pp 95–126

Kucera M, Weinelt M, Kiefer T, Pflaumann U, Hayes A, Weinelt M, Chen MT, Mix AC, Barrows TT, Cortijo E, Duprat J (2005) Reconstruction of sea-surface temperatures from assemblages of planktonic foraminifera: multi-technique approach based on geographically constrained calibration data sets and its application to glacial Atlantic and Pacific Oceans. Quaternary Science Reviews 24(7–9):951–998

Kumar R, Varandani D, Mehta BR, Singh VN, Wen Z, Feng X, Müllen K (2011) Fast response and recovery of hydrogen sensing in Pd–Pt nanoparticle–graphene composite layers. Nanotechnology 22(27):275719

Lau W (1995) Importation of baitworms and shipping seaweed: vectors for introduced species? In: Sloan D, Christensen M, Kelso D (eds) Environmental issues: from a local to a global perspective. Environmental Sciences Group Major, University of California, Berkeley, pp 21–38

Leao ZM, Kikuchi RK (2005) A relic coral fauna threatened by global changes and human activities, eastern Brazil. Mar Pollut Bull 51(5–7):599–611

Leão ZMAN, Kikuchi RKP, Ferreira BP, Neves EG, Sovierzoski HH, Oliveira MDM, Maida M, Correia MD, Johnsson R (2016) Brazilian coral reefs in a period of global change: a synthesis. Braz J Oceanogr 64:97–116

Lee YG, Jeong DU, Kang SR, Kim YW, Kim S, Jung EH, Lee JS (2012) The formation of hypoxia sediment and benthic foraminiferal change in Gamak Bay, southern coast of Korea. Ocean Polar Res 34:53–64. https://doi.org/10.4217/OPR.2012.34.1.053

Lee YG, Choi YH, Jeong DU, Lee JS, Kim YW, Park JJ, Choi JU (2016) Effect of abalone farming on seawater movement and benthic foraminiferal assemblage of Zostera marina in the inner bay of Wando, South Korea. Mar Pollut Bull 9:205–220

Lisiecki LE, Raymo ME (2005) A Pliocene-Pleistocene stack of 57 globally distributed benthic δ18O records. Paleoceanography 20(1)

Loeblich AR, Tappan H, Moore RC (1964) Treatise on invertebrate paleontology. University of Kansas Press

Loeblich AR Jr, Tappan H (1988) Foraminiferal genera and classification, vol 2. Van Nostrand Reinhold, New York, p 970

Lowe JJ, Walker MJC (1997) Temperature variations in NW Europe during the last glacial-interglacial transition (14-9^ 1^ 4C ka BP) based upon the analysis of coleopteran assemblages-the contribution of professor GR Coope. In: Quaternary Proceedings. Wiley, New York, pp 165–176

Lutze GF (1977) Neogene benthonic foraminifera from site 369, leg 41. In: Lancelot Y, Seibold E et al (eds) Initial reports of the Deep Sea drilling project, vol 41. U.S. Government Printing Office, Washington, DC, pp 659–666

Lutze GF (1979) Benthic foraminifers at site 397: faunal fluctuations and ranges in the quaternary. DSDP 47(1). https://doi.org/10.2973/dsdp.proc.47-1.111.1979

Lutze GF, Coulbourn WT (1984) Recent benthic foraminifera from the continental margin of Northwest Africa: community structure and distribution. Mar Micropaleontol 8(5):361–401

Machado AJ, Souza FBC (1994) Principais espécies de foraminíferos e Briozoários do Atol das Rocas. Rev Bras Geociênc 24(4):247–261

Magliocca A, Kutner AS (1965) Sedimentos de fundo daEnseada do Flamengo Ubatuba. Bol Inst Oceanogr 8:1–14

Mahiques MM, Duleba W, Sousa SHM (1995) Holocenic evolution of Flamengo Bay, Ubatuba, SP, Brazil. In: Book of abstracts of Regional Conference on Global Change, São Paulo, Dec. 4–6, pp 183–184

Marina Badaue (2017) Marina Badaue boat ride. http://marinabadaue.com.br/sobre-o-passeio

Martinez AS, Mendes LF, Leite TS (2012) Spatial distribution of epibenthic mollusks on a sandstone reef in the North east of Brazil. Braz J Biol 72:1–12

Matoba Y (1970) Distribution of recent shallow water foraminifera of Matsushima Bay, Miyagi prefecture, Northeast Japan. Sci RepTohoku Univ Sendai, Second Ser (Geology) 42:1–85

Matsushita S, Kitazato H (1990) Seasonality in the benthic foraminiferal community and the life history of *Trochammina hadai* Uchio in Hamana Lake, Japan. In: Paleoecology, biostratigraphy, paleoceanography and taxonomy of agglutinated foraminifera. Springer, Dordrecht, pp 695–715

Maxwell WGH (1968) Atlas of the Great Barrier Reef. Elsevier, Amsterdam, p 260

Maxwell WGH (1969) The structure and development of the great barrier reef. In: Hill D, Campbell KSW (eds) Stratigraphy and palaeontology: essays in honour of Dorothy Hill. Australian National University Press, Canberra, pp 195–217

Maxwell WGH, Swinchatt JP (1970) Great Barrier Reef: regional variation in a terrigenous–carbonate province. Geol Soc Am Bull 81:691–724

McGann M (2014a) Earliest record of the invasive foraminifera *T. hadai* in San Francisco Bay, California, USA. Mar Biodivers Rec 7:94–101

McGann M (2014b) Earliest record of the invasive foraminifera *T.* hadai in San Francisco Bay, California, USA. Mar Biodivers Rec 7(e94):1–7. https://doi.org/10.1017/S1755267214000888

McGann M, Sloan D (1996) Recent introduction of the foraminifer *T.hadai* Uchio in to san FranciscoBay, California, USA. Mar Micropaleontol 128:1–3

McGann M, Sloan D (1999) Benthic foraminifers in the regional monitoring Program's San Francisco estuary samples. In: 1997 annual report for the regional monitoring program for trace substances in the San Francisco estuary. San Francisco Estuary Institute, Richmond, CA, pp 249–258

McGann M, Sloan D, Cohen AN (2000) Invasion by a Japanese marine microorganism in western North America. Hydrobiologia 421:25–30

Medeiros JR (2009) Influência das águas da Bacia Hidrográfica Pirangi na balneabilidade das praias de Pirangi, nos municípios de Nísia Floresta e Parnamirim Rio Grande do Norte. Dissertação (Mestrado em Saneamento Ambiental; Meio Ambiente; Recursos Hídricos e Hidráulica), Universidade Federal do Rio Grande do Norte, Natal, Brasil, p 122

Medioli FS, Scott DB (1983) Holocene Arcellacea (thecamoebians) from eastern Canada (no. 21). Cushman Foundation for Foraminiferal Research, Washington, DC

Medioli FS, Scott DB (1988) Lacustrine thecamoebians (mainly arcellaceans) as potential tools for palaeo limnological interpretations. Palaeogeogr Palaeoclimatol Palaeoecol 62(1–4):361–386

Miranda LB (1972) Propriedades e variáveis físicas das águas da plataforma continental do Rio Grande do Sul. Tese, Instituto de Física, Universidade de São Paulo, São Paulo. [ Links ]

Miranda LB, Castro BM, Kjerfve B (2002) Principles of physical oceanography of estuaries. EDUSP, San Paulo. 424 p (academic 42)

Moraes SS, Machado AJ (2003) Avaliação das condições hidrodinâmicas de dois recifes costeiros do litoral norte do Estado da Bahia. Braz J Geol 33:201–210

Morey AE, Mix AC, Pisias NG (2005) Planktonic foraminiferal assemblages preserved in surface sediments correspond to multiple environment variables. Q Sci Rev 24(7–9):925–950

Morishima M, Chiji M (1951) Foraminiferal thanatocoenosesof Akkeshi Bay and its vicinity. Mem Coll Sci Univ Kyoto Ser B 10:113–117

Mortensen PB, Buhl-Mortensen L (2005) Coral habitats in the gully, a submarine canyon off Atlantic Canada. In: Freiwald A, Roberts JM (eds) Cold-water corals and ecosystems. Springer, Berlin, pp 247–277

Murray JW (1973) Distribution and ecology of living benthic Foraminiferids. Heinemann Educational Books, London, 274 p

Narayan YR, Pandolfi JM (2010) Benthic foraminiferal assemblages from Moreton Bay, south-East Queensland, Australia: applications in monitoring water and substrate quality in subtropical estuarine environments. Mar Pollut Bull 60:2062–2078

Natsir SM, Subkhan M (2011) The distribution of benthic foraminifera in coral reefs community and seagrass bad of Belitung Islands based on FORAM Index. Journal of coastal development 15(1):51–58.

Newman MC (1995) Quantitative methods in aquatic ecotoxicology. Lewis, Boca Raton

Odum EP (1988) Ecologia. Guanabara, Rio de Janeiro, 434 p

Oliveira-Silva P, Barbosa CF, Soares-Gomes A (2005) Distribution of macro benthic foraminiferaon Brazilian continental margin between 18°S–23°S. Rev Bras Geociênc 35:209–216

Oliveira-Silva P, Barbosa CF, Machado de Almeida C, Seoane JCS, Cordeiro RC, Turcq BJ, Soares-Gomes A (2012) Sedimentary geochemistry and foraminiferal assemblages in coral reef assessment of Abrolhos, Southwest Atlantic. Mar Micropaleontol 94:14–24

Patchineelam SR, Smoak JM (1999) Sediment accumulation rates along the inner eastern Brazilian continental shelf. Geo-Marine Letters 19(3):196–201

Patterson RT, Kumar A (2002) A review of current testate rhizopod (thecamoebian) research in Canada. Palaeogeogr Palaeoclimatol Palaeoecol 180(1–3):225–251

Peeters FJ, Acheson R, Brummer GJA, De Ruijter WP, Schneider RR, Ganssen GM, Ufkes E, Kroon D (2004) Vigorous exchange between the Indian and Atlantic oceans at the end of the past five glacial periods. Nature 430(7000):661–665

Pflaumann U (1975) Late quaternary stratigraphy based on planktonic foraminifera off Senegal. Meteor Forsch Ergebn 23:1–46

Phleger FB (1970) FORAMINIFERAL populations and marine MARSH processes 1. Limnol Oceanogr 15(4):522–534

Pimenta FM, Campos EJD, Miller JL, Piola AR (2005) A numerical study of the Plata River plume along the southeastern south American continental shelf. Braz J Oceanogr 53(3–4):129–146

Piola AR, Campos EJ, Möller OO Jr, Charo M, Martinez C (2000) Subtropical shelf front off eastern South America. J Geophys Res Oceans 105(C3):6565–6578

Piola AR, Matano RP, Palma ED, Möller OO Jr, Campos EJ (2005) The influence of the Plata River discharge on the western South Atlantic shelf. Geophys Res Lett 32(1)

Piola AR, Romero SI, Zajaczkovski U (2008) Space–time variability of the Plata plume inferred from ocean color. Cont Shelf Res 28(13):1556–1567

Pires Filho IA, Cycon DE (1987) Planning and managing's coastal resources. Braz Coast Manag 15:61–74

Radziejewska T, Gruszka P, Rokicka-Praxmajer J (2006) A home away from home: a meiobenthic assemblage in a ship's ballast water tank sediment. Oceanologia 48:259–265

Reich S, Di Martino E, Todd JA, Wesselingh FP, Renema W (2015) Indirect paleo-seagrass indicators (IPSIs): a review. Earth Sci Rev 143:161–186

Riveiros NV, Babalola AO, Boudreau RE, Patterson RT, Roe HM, Doherty C (2007) Modern distribution of salt marsh foraminifera and the camoebians in the Seymour–Belize inlet complex, British Columbia, Canada. Mar Geol 242(1–3):39–63

Rodrigues AR, Díaz TL, Pellizari VH (2014) Living foraminifera in a Brazilian subtropical coastal environment (Flamengo inlet, Ubatuba, São Paulo state–Brazil). In: Approaches to study living foraminifera. Springer, Tokyo, pp 195–227

Rohling E, Mayewski P, Abu-Zied R, Casford J, Hayes A (2002) Holocene atmosphere-ocean interactions: records from Greenland and the Aegean Sea. Climate Dynamics 18(7):587–593

Rosenthal Y, Holbourn A, Kulhanek D, Expedition 363 Scientists (2017) Expedition 363 preliminary report: western Pacific warm pool. In: Proceedings of the International Ocean Discovery Program, vol. 363, https://doi.org/10.14379/iodp.proc.363.101.2018

Rutherford S, D'Hondt S, Prell W (1999) Environmental controls on the geographic distribution of zooplankton diversity. Nature 400(6746):749–753

Rzehak A (1886) Uber foraminiferen: Verh. Naturf. Ver. Brunn (Brno), Sitzungsber, 24:8

Sanches TM, Kikuchi, RKP Eichler BB (1995) Ocorrência de foraminíferos recentes em Abrolhos, Bahia. Publicação Especial Instituto Oceanográfico, São Paulo, 11:37–47

Schafer CT, Winters GV, Scott DB, Pocklington P, Cole FE, Honig C (1995) Survey of living foraminifera and polychaete populations at some Canadian aquaculture sites; potential for impact mapping and monitoring. J Foraminifer Res 25(3):236–259

Schiebel R, Hemleben C (2005) Modern planktic foraminifera. Paläontol Z 79(1):135–148

Schiebel R, Hiller B, Hemleben C (1995) Impacts of storms on recent planktic foraminiferal test production and CaCO3 flux in the North Atlantic at 47 N, 20 W (JGOFS). Mar Micropaleontol 26(1–4):115–129

Schiebel R, Bijma J, Hemleben C (1997) Population dynamics of the planktic foraminifer *Globigerina bulloides* from the eastern North Atlantic. Deep Sea Res Pt 1 Oceanogr Res Pap 44(9):1701–1713

Schiebel R, Schmuker B, Alves M, Hemleben C (2002) Tracking the Recent and late Pleistocene Azores front by the distribution of planktic foraminifers. Journal of marine systems 37(1–3):213–227

Schmidt DN, Thierstein HR, Bollmann J (2004a) The evolutionary history of size variation of planktic foraminiferal assemblages in the Cenozoic. Palaeogeogr Palaeoclimatol Palaeoecol 212(1–2):159–180

Schmidt DN, Renaud S, Bollmann J, Schiebel R, Thierstein HR (2004b) Size distribution of Holocene planktic foraminifer assemblages: biogeography, ecology and adaptation. Mar Micropaleontol 50(3–4):319–338

Schmiedl G, Mackensen A, Müller PJ (1997) Recent benthic foraminifera from the eastern South Atlantic Ocean: dependence on food supply and water masses. Mar Micropaleontol 32(3):249–287

Schott W (1935) Die Foraminiferen in dem äquatorialen Teil des atlantischen Ozeans. Deutsche Atlantische Expedition 6:411–616

Schumacher S, Jorissen FJ, Mackensen A, Gooday AJ, Pays O (2010) Ontogenetic effects on stable carbon and oxygen isotopes in tests of live (Rose Bengal stained) benthic foraminifera from the Pakistan continental margin. Marine Micropaleontology 76(3–4):92–103

Schwager C (1866) Fossile Foraminiferen von Kar Nicobar, Reise der Österreichischen Fregatte Novara um Erde in den Jahren 1857, 1858, 1859 unter den Befehlen des Commodore B. Von Wüllerstorf-Urbair, Geologischer Theil, Geologische Beobachtungen, No. 2. Paläontologische Mittheilungen 2:187–268

Scott DB, Medioli FS, Schafer CT (2001) Monitoring of coastal environments using foraminifera and Thecamoebian indicators. Cambridge University Press, Cambridge, 176 p

Sen Gupta BK (1999) Foraminiferain marginal marine environments. In: Sen Gupta BK (ed) Modern Foraminifera. Kluwer, Dordrecht, pp 141–159

Sen Gupta BK, Turner RE, Rabalais NN (1996) Seasonal oxygen depletion in continental-shelf waters of Louisiana: historical record of benthic foraminifers. Geology 24:227–230

Setty MG, Nigam R (1982) Foraminiferal assemblages and organic carbon relationship in benthic marine ecosystem of western Indian continental shelf. Indian J Geo Mar Sci 11:225–232

Shackleton NJ, Opdyke ND (1973) Oxygen isotope and palaeomagnetic stratigraphy of equatorial Pacific core V28-238: oxygen isotope temperatures and ice volumes on a 10 5 year and 10 6 year scale. Quatern Res 3(1):39–55

Shank TM, Fornari DJ, Von Damm KL, Lilley MD, Haymon RM, Lutz RA (1998) Temporal and spatial patterns of biological community development at nascent deep-sea hydrothermal vents (9 50′ N, East Pacific Rise). Deep Sea Research Part II: Topical Studies in Oceanography 45(1–3):465–515

Smith LD, Wonham MJ, McCann LD, Reid DM, Carlton JT (1996) Shipping study II. Biological invasions by non indigenous species in United States waters: quantifying the role of ballast water and sediments. Parts I and II. U.S. Coast Guard, Report no. CG-D-02-97. Washington, DC, 97 pp and appendices

Souza CR, Vital H, Melo G Jr, Souza CR, da Silva Nogueira ML, Tabosa WF (2015) Sediment studies associated with drilling activity on a tropical shallow shelf. Environ Monit Assess 187:18. https://doi.org/10.1007/s10661-014-4232-5

Spindler M, Hemleben C, Bayer U, Bé AWH, Anderson OR (1979) Lunar periodicity of reproduction in the planktonic foraminifer Hastigerina pelagica. Marine Ecology Progress Series, pp 61–64

Stevenson MR, Dias-Brito D, Stech JL, Kampel M (1998) How do cold water biota arrive in a tropical bay near Rio de Janeiro, Brazil? Cont Shelf Res 18(13):1595–1612

Strotz LC Brock GA (2006) Holocene foraminifera from the reefs surrounding Viti Levu, Fiji Islands: A test of the FORAM Index. Anuário do Instituto de Geociências 29(1):305–306

Suess E (1980) Particulate organic carbon flux in the oceans—surface productivity and oxygen utilization. Nature 288(5788):260–263

Swindles GT, De Vleeschouwer F, Plunkett G (2010) Dating peat profiles using tephra: stratigraphy, geochemistry and chronology. Mires and Peat 7

Testa V, Bosence DWJ (1999) Physical and biological controls on the formation of carbonate and siliciclastic bedforms on the northeast Brazilian shelf. Sedimentology 46:279–301

Thiede J (1975) Distribution of foraminifera in surface waters of a coastal upwelling area. Nature 253(5494):712–714

Thomas E, Gapotchenko T, Varekamp JC, Mecray EL, Ten Brink MB (2000) Benthic foraminifera and environmental changes in Long Island sound. J Coast Res 16:641–655

Toyoda K, Kitazato H (1995) Paleoenvironmental changes of Yokohama port since 1870 based on benthic foraminiferal fossils. Annu Rep Yokohama Env Res Inst 116:11–26

Tunnicliffe V, Embley RW, Holden JF, Butterfield DA, Massoth GJ, Juniper SK (1997) Biological colonization of new hydrothermal vents following an eruption on Juan de Fuca Ridge. Deep Sea Research Part I: Oceanographic Research Papers 44(9–10):1627–1644

Tyson RV (1995) Sedimentary organic matter: organic facies and palynofacies. Chapman and Hall, Londres, 615 p

Ueki I, Kashino Y, Kuroda Y (2003) Observation of current variations off the New Guinea coast including the 1997–1998 El Niño period and their relationship with Sverdrup transport. Journal of Geophysical Research: Oceans, 108(C7)

Uthicke S, Nobes K (2008) Benthic foraminifera as ecological indicators for water quality on the great barrier reef. Estuar Coast Shelf Sci 78:763–773

Uthicke S, Thompson A, Schaffelke B (2010) Effectiveness of benthic foraminiferal and coral assemblages as water quality indicators on inshore reefs of the great barrier reef, Australia. Coral Reefs 29:209–225

Van Hengstum PJ, Reinhardt EG, Boyce JI, Clark C (2007) Changing sedimentation patterns due to historical land-use change in Frenchman's bay, Pickering, Canada: evidence from high-resolution textural analysis. J Paleo 37(4):603–618

Van Morkhoven FP, Berggren WA, Edwards AS, Oertli HJ (1986) Cenozoic cosmopolitan deep-water benthic foraminifera, vol 11. Elf Aquitaine, Pau

Velásquez J, López-Angarita J, Sánchez JA (2011) Benthic foraminifera as indicators of ecosystem resilience in protected and non-protected coral reefs of the Southern Caribbean. Biodivers Conserv 20:3591–3603. https://doi.org/10.1007/s10531-011-0152-7

Venera-Ponton DE, Diaz-Pulido G, McCook LJ, RangelCampo A (2011) Macroalgae reduce growth of juvenile corals but protect them from parrotfish damage. Mar Ecol 421:109–115

Vital H, Furtado SFL, Gomes MP (2010a) Response of Apodi-Mossoro estuary-Incised Valley system (ne Brazil) to sea-level fluctuations. Braz J Oceanogr 58:13–24

Vital H, Gomes MP, Tabosa WF, Frazao EP, Santos CLA, Placido JS Jr (2010b) Characterization of the Brazilian continental shelf adjacent to Rio Grande do Norte state, NE Brazil. Braz J Oceanogr 58:43–54

Vroom PS, Braun CL (2010) Benthic composition of a healthy subtropical reef: baseline species-level cover, with an emphasis on algae, in the northwestern Hawaiian islands. PLoS One 5(3):e9733. https://doi.org/10.1371/journal.pone.0009733

Vroom PS, Page KN, Kenyon JC, Brainard RE (2006) Algae-dominated reefs. Am Sci 94:430–437

Webster JM, Wallace L, Silver E, Applegate B, Potts D, Braga JC, Riker-Coleman K, Gallup C (2004) Drowned carbonate platforms in the Huon Gulf, Papua New Guinea. Geochemistry Geophysics Geosystems 5(11)

Whittaker RJ, Triantis KA Ladle RJ (2010) A general dynamic theory of oceanic island biogeography: extending the MacArthur–Wilson theory to accommodate the rise and fall of volcanic islands. The theory of island biogeography revisited 88:115

Williams RJ, Griffiths FB, Van Der Wal EJ, Kelly J (1988) Cargo vessel ballast water as a vector for the trans-port of non-indigenous marine species. Estuar Coast Shelf Sci 26:409–420

Wilson B, Ramsook A (2007) Population densities and diversities of epiphytal foraminifera on nearshore substrates, Nevis, West Indies. J Foraminifer Res 37:213–222

Woodland WA, Charman DJ, Sims PC (1998) Quantitative estimates of water tables and soil moisture in Holocene peatlands from testate amoebae. Holocene 8(3):261–273

Woodroffe SA, Horton BP, Larcombe P, Whittaker JE (2005) Intertidal mangrove foraminifera from the central Great Barrier Reef shelf, Australia: implications for sea-level reconstruction. J Foraminifer Res 35(3):259–270. https://doi.org/10.2113/35.3.259

Wulff JL (2006) Ecological interactions of marine sponges. Can J Zool 84:146–166

Yanko V, Kronfeld J, Flexer A (1994) Response of benthic foraminifera to various pollution sources: implications for pollution monitoring. J Foraminifer Res 24:1–17

Yanko V, Arnold AJ, Parker WC (1999) Effects of marine pollution on benthic foraminifera. In: Modern foraminifera. Springer, Dordrecht, pp 217–235

Zachos JC, Shackleton NJ, Revenaugh JS, Pälike H, Flower BP (2001) Climate response to orbital forcing across the Oligocene-Miocene boundary. Science 292(5515):274–278

Zaninetti L, Bronnimann P, Beurlen G, Moura JA (1977a) La mangrove de Guaratiba et la Baia de Sepetiba, etat de Rio de Janeiro, Brésil: Foraminifères et écologie. Note preliminaire. Compte rendu des séances, PS Histoire Naturelle, Géneve 11(1–3):39–44

Zaninetti L, Brönnimann P, Beurlen G, Moura JA (1977b) La mangrove de Guaratiba et la Baie de Sepetiba. Etat de Rio de Janeiro, Brésil: Foraminiféres et écologie. Arch Sci Genève 30(2):161–178

Zar JH (1984) Biostatistical analysis, 2nd edn. Prentice-Hall, Englewood Cliffs, NJ. 718 p

Zaric S, Donner B, Fischer G, Mulitza S, Wefer G (2005) Sensitivity of planktic foraminifera to sea surface temperature and export production as derived from sediment trap data. Mar Micropaleontol 55(1–2):75–105

# Index





P. P. Beck Eichler, C. P. Barker, *Benthic Foraminiferal Ecology*,
https://doi.org/10.1007/978-3-030-61463-8

The manufacturer's authorised representative in the EU is Springer Nature Customer Service Centre GmbH, Europaplatz 3, 69115 Heidelberg, Germany. If you have any concerns regarding our products, please contact ProductSafety@springernature.com

Printed and bound by CPI Group (UK) Ltd, Croydon, CR0 4YY
28/09/2026
02224086-0003